Contemporary Issues in Science Communication

Series editor: **Clare Wilkinson**,
University of the West of England, Bristol

As science communication continues to establish itself as a discipline in the 21st century, this book series links the present with the past to develop our understanding of its history, both in practice and as an academic discipline.

Forthcoming in the series:

*Race and Socio-Cultural Inclusion in Science Communication;
Innovation, Decolonisation and Transformation*
Edited by **Elizabeth Rasekoala**

Find out more at
bristoluniversitypress.co.uk/
contemporary-issues-in-science-communication

International advisory board:

Alison Anderson, University of Plymouth, UK
Ayman Elsayed, Planetarium Science Centre, Library of Alexandria, Egypt
Birte Fahnrich, Berlin-Brandenburg Academy of Sciences and Humanities, Germany
Carla Almeida Museu da Vida, Fundação Oswaldo Cruz, Rio de Janeiro, Brasil
Daria Denisova, Information Technologies, Mechanics and Optics University, Russia
Elizabeth Rasekoala, African Gong, South Africa
Emily Dawson, University College London, UK
Erik Stengler, The State University of New York College at Oneonta, US
Jenny Martin, University of Melbourne, Australia
Lesley Markham, Center for Open Science, US
Marina Joubert, University of Stellenbosch, South Africa
Padraig Murphy, Dublin City University, Ireland

Find out more at
bristoluniversitypress.co.uk/
contemporary-issues-in-science-communication

CONTEMPORARY ISSUES IN SCIENCE COMMUNICATION

QUEERING SCIENCE COMMUNICATION

Representations, Theory, and Practice

Edited by
Lindy A. Orthia and Tara Roberson

First published in Great Britain in 2025 by

Bristol University Press
University of Bristol
1-9 Old Park Hill
Bristol
BS2 8BB
UK
t: +44 (0)117 374 6645
e: bup-info@bristol.ac.uk

Details of international sales and distribution partners are available at bristoluniversitypress.co.uk

© Bristol University Press 2025

British Library Cataloguing in Publication Data
A catalogue record for this book is available from the British Library

ISBN 978-1-5292-2440-5 hardcover
ISBN 978-1-5292-2441-2 paperback
ISBN 978-1-5292-2442-9 ePub
ISBN 978-1-5292-2443-6 ePdf

The right of Lindy A. Orthia and Tara Roberson to be identified as editors of this work has been asserted by them in accordance with the Copyright, Designs and Patents Act 1988.

All rights reserved: no part of this publication may be reproduced, stored in a retrieval system, or transmitted in any form or by any means, electronic, mechanical, photocopying, recording, or otherwise without the prior permission of Bristol University Press.

Every reasonable effort has been made to obtain permission to reproduce copyrighted material. If, however, anyone knows of an oversight, please contact the publisher.

The statements and opinions contained within this publication are solely those of the editors and contributors and not of the University of Bristol or Bristol University Press. The University of Bristol and Bristol University Press disclaim responsibility for any injury to persons or property resulting from any material published in this publication.

Bristol University Press works to counter discrimination on grounds of gender, race, disability, age and sexuality.

Cover design: Liam Roberts
Front cover image: Lan Zhang

Contents

Series Editor Preface	viii
Notes on Contributors	x
Acknowledgements	xxi
Terminology and Sensitive Content in This Book	xxii
Introduction *Tara Roberson and Lindy A. Orthia*	1

PART I Negotiating Queer Identities with Science, Technology, and Medicine

1. Where to 'Keep' the Queer: Contestations and Anxieties in Clinical Communications — 17
   Aritra Chatterjee

   Practice Spotlight: Gender and Sex in Research Communications — 29
   Sophia Frentz

2. The Question of Queer Complexity: Science Communication and Queer Activism — 32
   V de Kauwe and Emily Standen

3. Queer Interests in Technology and Innovation Discourse — 48
   Tara Roberson

   Practice Spotlight: All We Need Is … The Endosymbiotic Love Calendar — 60
   Annalaura Alifuoco, Natalie E.R. Beveridge, Yasmine Kumordzi, and Hwa Young Jung

   Practice Spotlight: GENDERS: Shaping and Breaking the Binary, an Exhibition at Science Gallery London — 64
   Helen Kaplinsky and Jessie Krish

   Teaching Notes for Part I — 67

PART II Representations of Queerness in Public Science Communication

4   Queering Science Museums, Science Centres, and Other Public Science Institutions   71
Eleanor S. Armstrong and Simon J. Lock

Practice Spotlight: Queer by Nature: The LGBTQ+ Natural History Tour   82
Josh Davis

Practice Spotlight: Science Queers: Overacted Representation in Science Communication   84
Òscar Aznar-Alemany

Practice Spotlight: Science is a Drag! Online Events   87
Carla Suciu, Brynley Pearlstone, and Sam Langford

5   Queer Characters in Science-themed Fiction   89
Lindy A. Orthia and Leo P. Visser

Practice Spotlight: Using #QueerInSTEM and Related Hashtags to Promote Your Science Communication   101
Luis Lopez and Alberto I. Roca

Practice Spotlight: Queer Science Blogs: Public Communication Before the Age of Social Media   104
Ron Buckmire and Alberto I. Roca

Teaching Notes for Part II   107

PART III Queer People in Science Communication Communities

6   *Malayang Paglaladlad para sa Mapagpalayang Paglalahad*: Coming Out and Queering Science Communication in Contested Spaces   111
John Noel Viaña, Mario Carlo Severo, Miguel Barretto-Garcia, Paul James Magtaan, Jason Tan Liwag, Roemel Jeusep Bueno, Christer de Silva, and Ruby Shaira Panela

Practice Spotlight: Queer Scientists PH: Visibility Towards Community Building and Empowerment   124
Jason Tan Liwag, Jay S. Fidelino, Rey Audie S. Escosio, Almira B. Ocampo, and Nikki Santos-Ocampo

Practice Spotlight: 500 Queer Scientists at the Sydney Gay and Lesbian Mardi Gras   127
Alice Motion and Hervé Sauquet

7	Including Queerness and Improving Belonging of Intersectional Queer Identities in Science Communication Communities *Katherine Canfield*	129
	Practice Spotlight: Rainbow Spectrums: Embracing Our Queer Disabled Family in Science Communication *V de Kauwe and Kai Fisher*	141
8	Have Rainbow, Will Collect Data: How Citizen and Community Science Engages Queer Volunteers *Todd A. Harwell*	144
	Teaching Notes for Part III	156

PART IV Queering Institutional Science Communication Agendas

9	Science OUTreach: A Queer Approach to Science Communication Practice *Alice Motion and Lee Wallace*	161
	Practice Spotlight: Queer Communicators in Environmental, Climate Change, and Sustainability Conversations *Franzisca Weder*	172
	Practice Spotlight: How LGBTIQA+ Representation in Organization Leadership Impacts Inclusivity and Visibility *Sarah Durcan and Andrea Bandelli*	175
	Practice Spotlight: Outer Edge: Queer(y)ing STEM Collections – A Community Workshop *Eleanor S. Armstrong and Sophie Gerber*	178
10	The Possibilities of Queer in Science Communication Teaching and Pedagogies *Simon J. Lock and Eleanor S. Armstrong*	181
11	Queering Science Communication Theory Beyond Deficit and Dialogue Binaries *Lindy A. Orthia and V de Kauwe*	191
	Teaching Notes for Part IV	204

Conclusions 206
Tara Roberson and Lindy A. Orthia

Index 213

Series Editor Preface

Clare Wilkinson

As science communication continues to establish itself as a discipline in the 21st century, there has never been a better time to consider contemporary science communication and its practices. Misinformation and digital marketing are changing the context for science journalism. Emerging political eras are altering the way we think about expertise and trust in policymaking, as well as the power of protest. Predatory publishing, open access and the use of social media are presenting novel contexts for researchers to consider in the communication of their work. Meanwhile, there are contemporary scientific and technological developments that are consistently generating important ethical and social questions. Against this backdrop in 2020, we faced one of the greatest global health issues in a generation, with the onset of the COVID-19 pandemic. Whilst international events, and social movements have drawn heightened attention to questions of inclusion, equity and the abuse of power.

Contemporary Issues in Science Communication is a multidisciplinary series, welcoming submissions from a wide range of disciplinary areas, and international in outlook. Books contained in the series seek to be engaging, straightforward and conversational style, and of interest to science communication practitioners as well as academic audiences.

Books published in the Contemporary Issues in Science Communication series consider such science matters and their relationships to communication, engagement and broader social conversations. The series links the present with the past by publishing titles that develop our understanding of the history of science communication, both in practice and as an academic discipline. However, they also cover a range of topics relevant to contemporary science communication, including, but not limited to: definitions, history and ethics of science communication; expertise, replication and trust; interdisciplinary knowledge; ideologies; knowledge and new forms of media; public policy; gaming, sci-art and visual communication and inclusivity in science communication. The scope of the series is broad but so are the challenges facing science communicators and public engagement practitioners.

SERIES EDITOR PREFACE

In this book, the first in the series, Tara Roberson and Lindy A. Orthia encapsulate these ambitions. In *Queering Science Communication: Representations, Theory and Practice*, they bring together an edited collection of insights that highlight the intersections of queerness and science communication. These are, as the editors point out, insights that have been lacking in the field of science communication and are much needed. Their authors reflect a diversity of perspectives from LGBTIQA+ people that are truly international in scope. These perspectives draw on a wide range of examples, both theoretical and practical to tackle the question 'what does "queering science communication" mean?' with both authenticity and ownership.

Tara and Lindy have created a home for both research focused accounts and practice informed spotlights, as well as providing useful teaching notes, activities and prompts for those who may wish to embrace this text in their classrooms, organizations, science centres, museums, or other science communication settings. There is also a lot of heart in this book. Experiences and examples that I know will have been challenging for some of the authors to share. Provocations and calls to action for all those associated with science communication to reflect on. These insights sit alongside moments of humour, bringing joy and light from a vibrant and growing LGBTIQA+ science communication community/ies, which are so willingly shared with readers. The authors handle the challenging issues raised in this book with sensitivity, specificity and grace.

I hope you will find this book as interesting, challenging, but also as enjoyable, as I did to read. If you are interested in becoming involved in the series personally, please do get in touch. Finally, thank you Lindy and Tara for providing the first contribution to the series, and for trusting us with this essential and important collection.

Notes on Contributors

Annalaura Alifuoco (they/them) works across writing and performance and academia. Their interdisciplinary practice combines experimentation, art making, posthuman ecologies, and queer theory to reframe the human within entangled ecologies. Themes of sexuality, performativity and community are explored through art, science, new materialism, and social justice. Through this work, aspects of queerness – such as sexuality, gender, and kinship – are woven together to form heterotopic alliances and new ethical ontologies.

Eleanor S. Armstrong (she/her) is a postdoctoral researcher at Stockholm University (Sweden), taking a critically queer feminist approach to informal science education, and the cultural geographies of space science. Her previous affiliations involved research at the University of Delaware (USA) and the University of Cambridge (UK); and writing her doctorate at University College London (UK). Armstrong developed *Queering the Science Museum* at the Science Museum London and *Bridging Binaries* programme at Whipple Museum for the History of Science and the Sedgewick Earth Science Museum as tours that explore the queer potentials in informal education spaces. She co-organizes *Space Science in Context*, which was noted in NASA decadal reviews for its gender inclusive conference practices.

Òscar Aznar-Alemany (he/him) is a Chemistry PhD, translator (Catalan, Spanish, English, Portuguese) and science communicator who teaches introduction to translation and specialized scientific and technical translation at the Universitat Autònoma de Barcelona and edits a trilingual chemistry dictionary published online by the TERMCAT. Òscar has researched pyrethroid pesticides and flame retardants in sea life at the CSIC (Barcelona) with two internships at the Ministry of the Environment, Conservation and Parks (MECP) (Toronto) and performs science shows at a wide range of events, from science conferences to nightclubs, generally as the drag queen Lana Vuli (she/her). Lana Vuli is also the hostess of the YouTube channel Science Queers.

Andrea Bandelli (he/him) is Head of International Relations at Vrije Universiteit Amsterdam. Andrea's academic background includes Masters' degrees in Economics and Science Communication and a PhD in Social Sciences, specializing in scientific citizenship. Andrea is a member of the Munich-based Deutsches Museum board, of the scientific council at Universcience in Paris, and from 2016 to 2022 he was executive director at Science Gallery International. He has worked for science museums, government organizations, and universities across Europe, the US, South Africa and Brazil, leading innovative projects on science, art, democracy, and public participation.

Miguel Barretto-Garcia (they/them) is a poet, and they earned their PhD in decision neuroscience at the University of Zurich. Their research seeks to understand the neuro-computational intersections between perceptual and economic decision making in humans using a combination of neuroimaging, computational modelling, and non-invasive brain stimulation. Some of their literary works appear in *Rattle*, *wildness*, *harana poetry*, *The Quarterly Literary Review Singapore*, and *Cordite Poetry Review* in Australia. ORCID: Miguel Barretto Garcia (0000-0001-9054-7859). Twitter: @miguelabgarcia.

Natalie E.R. Beveridge (she/her) works with care as an NHS physician. Previously Natalie was an infectious diseases immunologist researching the interaction between the human immune system and our microbiological exterior. Nowadays she prefers to consider health at the wider intersections of knowledge production, ethics, society, philosophy, nutrition, sustainability, ecology, and permaculture.

Ron Buckmire (he/him) is Professor of Mathematics and former Associate Dean for Curricular Affairs at Occidental College. His academic interests include applied mathematics, mathematical modelling, numerical analysis, mathematics education, LGBT history, and sexual orientation law. Among many service activities, Buckmire created the Queer Resources Directory at www.qrd.org in 1991 and is co-founder of Spectra, the association for LGBT mathematicians and their allies. In 2011, he received the Educator of the Year award from the National Organization of Gay and Lesbian Scientists and Technical Professionals (NOGLSTP). Buckmire has been blogging at The Mad Professah Lectures since 2003 at madprofessah.com.

Roemel Jeusep Bueno (he/him) is a neuroscience and cell biology researcher at Charité – Universitätsmedizin Berlin. He is currently doing his PhD under the Berlin-Brandenburg School for Regenerative Therapies and Humboldt Universität zu Berlin. His main work focuses on the role of human extracellular matrix proteins in central nervous system regeneration. He is a staunch activist against the ongoing culture of fascism in the Philippines and advocates to make healthcare and education more accessible for underprivileged queer Filipinos. ORCID: Roemel Jeusep Bueno (0000-0001-9780-1485).

Katherine Canfield (she/her) is a queer, US-based, interdisciplinary researcher-practitioner with training and expertise in environmental and coastal social science, environmental and social justice, and science communication. Her doctoral research used critical qualitative social science to investigate the justice of tourism development on Catalina Island, California. She is most excited about work that integrates social justice and equity with STEMM and science communication, as exhibited in her past research with the Metcalf Institute on inclusive science communication. She currently works as a postdoctoral translational scientist with the US Environmental Protection Agency.

Aritra Chatterjee (she/they) self-identifies as trans-feminine and queer and is an MPhil Trainee in Clinical Psychology at University of Calcutta, Kolkata, India. Their express interest lies in LGBTQIA+ mental health, native queer subcultures involving *kothi-hijra-dhurani* communities, and ways of queering praxis in mental healthcare. Their Masters' thesis focused on understanding the nature of interactional experiences that LGBT+ clients based in Kolkata, India have with mental health professionals in contexts of care. They have authored several articles on queer mental health and have been invited as speaker for multiple panel discussions touching important elements of such conversations.

Josh Davis (he/him) graduated with a degree in biology and worked briefly in animal conservation before moving into science journalism eight years ago. During this time, he has worked for and had pieces published in the BBC, *The Guardian*, *The Times* and Mongabay. For the past few years Josh has been a science writer for the Natural History Museum, London, where he writes about the research conducted by the 300 scientists who work at the museum. During this time, he helped develop and lead the museum's first LGBTQ+ tour, exploring the history of queer animals and people from a natural history perspective.

V de Kauwe (she/her) is the founder and current head of disability organization Science Alliance. She is a Person with Disabilities herself, and her PhD research created and evaluated science-based education programmes for people with intellectual disabilities. V has published on postcolonialism and third-world issues and allegorical portrayals of marginalized people in science fiction. Her publications also extend into the societal constructs of gender and disability, as well as the philosophy of politics and social justice.

NOTES ON CONTRIBUTORS

Christer de Silva (he/him) is a science communicator at The Mind Museum in the Philippines where he develops educational programmes, facilitates dialogues, designs exhibits, and performs science shows. He explores how science can bring social justice through research, outreach, and education. Christer is also an Assistant Professorial Lecturer at the Department of Communication at the De La Salle University Manila and a Lecturer at the Ateneo de Manila University. ORCID: Christer de Silva (0000-0003-4782-339X).

Sarah Durcan (she/her) is the Acting Executive Director of Science Gallery International. She has degrees in Communications from Dublin City University and Cultural Policy and Arts Management from University College Dublin. She has served on the boards of the Abbey Theatre, Theatre Forum and GAZE Film Festival, is on the advisory panel of Dublin City Council's Culture Company, and is a member of the Expert Network of the World Economic Forum. She was a lead organizer of the #WakingTheFeminists campaign to achieve gender equality in Irish theatre. She has produced and consulted for The Corn Exchange, Dublin Theatre Festival, Dublin Fringe Festival, ARCANE Collective Dance Company, and Theatre Lovett.

Rey Audie S. Escosio (he/him) works as a researcher of the Modelling and Applications Group of the Institute of Mathematics, University of the Philippines Diliman for its COVID-19 mathematical modelling project. He recently finished his MS Applied Mathematics degree from the same institute and graduated from the same university with a degree in BS Applied Physics. For both of his studies, he is a recipient of the government's Department of Science and Technology scholarships. His graduate research – gradient-based optimization and its applications – was awarded a student poster prize from the Society of Mathematical Biology 2020 Annual Meeting. His research interests include mathematical biology, optimization, and complex systems. ORCID: Rey Audie Escosio (0000-0001-6802-017X). Twitter: @rsescosio.

Jay S. Fidelino (he/him) is a researcher at the Biodiversity Research Laboratory of the Institute of Biology, University of the Philippines Diliman. He recently finished his MS Biology degree from the same institution. He works on the ecology and biogeography of Philippine wildlife, restoration ecology, and biodiversity science communication. He has authored peer-reviewed publications on the ecology and responses of plants, birds, and bats to disturbances in Philippine forests. He is also a member of the IUCN SSC Small Mammal Specialist Group, the Board of Trustees of the Biodiversity Conservation Society of the Philippines, and Advocates of Science and

Technology for the People (AGHAM). ORCID: Jay Fidelino (0000-0002-0734-9286). Twitter/Instagram: @jsfidelino.

Kai Fisher (he/they) is a working member of disability organization Science Alliance. He has a passion for using visual arts, drama, and music for disability education and empowerment. They are currently Science Alliance and Science Pirates production manager and gender diversity representative. He is the Games-master for ACT Down Syndrome Association.

Sophia Frentz (they/them) has a PhD in medical genetics from the University of Melbourne and now works in data, aiming to enable data driven decision making while keeping your information as safe as possible. They have published writing in *Archer*, *The Conversation*, and *Lateral Magazine*, and spoken at conferences across Australia and New Zealand. Sophia was an Out for Australia 30 under 30 LGBTQIA+ role model for 2020, and currently sits on the board of Out for Australia. Sophia is non-binary, disabled, and neurodiverse, and their work aims to create a better world for people both like and unlike them.

Sophie Gerber (she/her) studied history and cultural and social anthropology at the University of Vienna. She gained her PhD in the history of technology at the Technical University of Munich in 2014. She was involved in exhibitions at Deutsches Museum Munich, the University of Vienna and the House of Austrian History. She has been a custodian for household technology and food since 2019 and leads the 'Focus Gender' venture at the Technical Museum Vienna. Her research interests include collections research and material culture, the history of technology and consumption in the 20th and 21st centuries, gender and queer studies.

Todd A. Harwell (he/him) is a Teaching Assistant Professor with the University Honors College at Portland State University. He holds a PhD in Environmental Sciences from Oregon State University and has been affiliated with the Center for Community and Citizen Science at the University of California, Davis and the Oregon Museum of Science and Industry (OMSI). Todd is a former marine scientist turned environmental educator turned environmental social scientist who explores the relationships between science engagement opportunities (such as science communication, informal STEM education, and citizen and community science) and the LGBTQ+ community. Twitter: @ToddAHarwell.

Hwa Young Jung (she/her) is a multidisciplinary artist working in the arts, culture and sciences, facilitating collaborative projects and workshops. She works with people to co-create projects, often using games and play to explore social issues. Based in the northwest of England she has been producing work with a range of people (young people in libraries, care workers, freshwater biologists, and criminologists) in England and internationally for almost ten years.

Helen Kaplinsky (she/her) is a curator and writer based in Helsinki and London. Her most recent exhibition is GENDERS, Science Gallery London (2020). She is currently undertaking a PhD in Curating at Liverpool John Moores University, Exhibition Research Lab and examining virtual conditions from a feminist perspective as part of the research collaboration Beyond Matter. Her research spans cyberfeminist legacies, postdigital identity and ownership. In 2015 she co-founded Res., workspace/gallery (London), see publication *Alembic* (co-edited and co-authored, AND, 2018).

Jessie Krish (she/her) is a cultural worker based in London. Her interdisciplinary research explores conflicts that shape the civic space, involving collaboration with non-artists in and beyond a gallery context. Recently, Jessie commissioned *Net-Zero Youth Voice* in partnership with Imperial College London's Environmental Research Group and Cove Park artist residency centre (2022); curated film and poster series *Cultural Field*, Turf Projects (2022); co-edited e-flux Reader 'Loot and Looting' (2020), and produced *GENDERS: Shaping and Breaking the Binary*, Science Gallery London (2020). She is the Gallery Manager at Cell Project Space.

Yasmine Kumordzi (she/her) is a PhD student at the chemistry department in Durham University, UK researching the drug target validation of sphingosine kinase in *Leishmania mexicana* parasites. She is passionate about using creativity and art to engage the public and create a safe encouraging space in which science can be explored by all. In recent years she has taken part in 'para-site-seeing' (para-site-seeing.org) and conceptualized 'Endosymbiotic Love' (endolove.slyrabbit.net), both sci-art projects exploring travel and relationships between humans and microorganisms. Through both of these projects she has an understanding of interdisciplinary science communication with contents outside of heteronormative science communication exploring diverse gender and sexuality.

Sam Langford (he/him) is a freelance science communicator and public engagement professional with over nine years of experience in science communication. From whisky tasting up mountains to establishing a science festival, interviewing NASA astronauts to sending a Tunnock's Teacake into the upper atmosphere, he has a quite unique background. As part of the Glasgow-based team of LGBTQ+ science communicators behind Science is a Drag, he has brought together his love of combining science, arts, and queer performance.

Jason Tan Liwag (he/him) is an actor, film critic, and writer completing his MS in Molecular Biology and Biotechnology at the University of the Philippines Diliman. His research explores the effects of chronic endocrine disruptor exposure on breast cancer progression. He has written about science, film, culture, and LGBTQIA+ issues for CNN Philippines Life and Rappler, among other publications. He is the founder and president of Queer Scientists PH – an organization highlighting the narratives of queer Filipino scientists globally. He was honoured as one of Attitude Magazine's 101 LGBTQ Trailblazers Changing the World Today. ORCID: Jason Tan Liwag (0000-0003-0661-302X). Twitter: @jaseybel.

Simon J. Lock (he/they) is an Associate Professor in Science Communication and Governance in the Department of Science and Technology Studies at University College London (UCL). They are currently Director of qUCL, UCL's research centre on gender and sexual diversity and have been on the organizing committee for the Public Understanding of Science Seminar Series at London School of Economics/UCL since 2002. Their research focuses on science in public cultures, science technology and gender, sexualities and race, the governance of science, and the sociology of new technologies.

Luis Lopez (he/him) is a graduate student in sociology at Indiana University. Previously, he was an intramural fellow in the National Human Genome Research Institute at the National Institutes of Health. Luis is interested in social science methodology. Lopez received his bachelor's degree in Peace and Conflict Studies from Westminster College. His LGBT advocacy includes work with LGBT youth at the Utah Pride Center where he received the Community Icon Award. ORCID: Luis Alberto Lopez (0000-0002-8590-0782).

Paul James D. Magtaan (she/her), known as Paige, is a public school science teacher at Kapt. Jose Cardones Integrated School. She holds a Master of Arts in Education, Major in Educational Management from Pasig Catholic College. She focuses on how students can learn science effectively. Her goal is to create safe spaces for LGBTQ+ students inside and outside the classroom.

Alice Motion (they/she) is an Associate Professor who leads the Science Communication, Outreach, Participation and Education (SCOPE) Group at the School of Chemistry, University of Sydney. Alice's research focuses on open science and SCOPE with an overarching theme of connecting people with science through research and practice. Alice was awarded the 2020 Eureka Prize for Promoting Understanding of Science, hosts a weekly science segment on FBi Radio, and is the co-host of ABC (Australian Broadcasting Corporation) podcast Dear Science.

Almira B. Ocampo (she/her) is currently an MS Chemistry student from the University of the Philippines Diliman. She is also a researcher of the Virtual Biochemical Explorations Laboratory at the Institute of Chemistry, UP Diliman. She graduated BS Chemistry from the same university as a government's Department of Science and Technology scholar. Almira's research interests include computational biochemical systems such as computer-aided drug design, computational agrochemicals design, and in silico toxicity analysis. She has authored a peer-reviewed publication on monoterpenes as a potential biopesticide for rice weevils. Her current research is on agrochemicals design for pests in a computational and theoretical approach and in silico toxicity analysis of natural products. Twitter: @abocampo_.

Lindy A. Orthia (she/her) is an Honorary Senior Lecturer at the Australian National University School of Sociology. Until 2021 she was a Senior Lecturer in Science Communication at the same university. She has authored over 40 peer-reviewed papers and chapters and edited two books prior to this one: *Doctor Who and Race* (2013, Intellect) and *Doctor Who and Science: Essays on Ideas, Identities and Ideologies in the Series* (2020, McFarland, edited with Marcus K. Harmes). Lindy was a political activist in Melbourne, Australia for the duration of the 1990s, active in many queer campaigns. She now works for a non-governmental organization as a researcher and communicator.

Ruby Shaira Panela (she/her) is an independent science writer currently serving as Media Relations Specialist at the International Rice Research Institute (IRRI), a non-profit organization focused on agricultural research for rice-based economies. Her articles have been published in various news outlets like *Rappler, Nature Index, Mongabay* and *Asian Scientist Magazine*. She also teaches science journalism at the University of the Philippines Los Baños. Her advocacy is focused on promoting science and technology as an essential contributor towards economic development.

Brynley Pearlstone (he/him) is a triple threat: astrophysicist, science communicator, and healthcare worker, based in Glasgow. His science communication work focuses on spotlighting queer scientists for public audiences. He believes that LGBTQ+ issues belong beyond the pigeonhole of diversity but are relevant to everyone.

Tara Roberson (she/her) is a queer science communicator and social scientist. As a research fellow, Tara works with quantum physicists to understand the implications of emerging technologies in their field at the ARC Centre of Excellence for Engineered Quantum System. She has authored peer-reviewed publications on hype in science communication, queering science communication, and responsible development of emerging technologies. Tara is also a trained facilitator for University of Queensland LGBTIQA+ Ally workshops and delivers content on diversity and inclusion for queer people.

Alberto I. Roca (he/him) is Executive Director of US-based non-profit DiverseScholar whose mission is to diversify the STEM workforce including through *DiverseScholar* magazine and the SciCommDiversity.org Fellowship that awards travel funds to attend science journalism/communication conferences. While a biochemistry postdoc, he created the career portal MinorityPostdoc.org, founded the Postdoc Committee of the Society for the Advancement of Chicanos and Native Americans in Science (SACNAS), and co-founded the Diversity Committee of the National Postdoctoral Association. As an ally, Roca has mentored LGBT students through STEM doctoral education and sponsored organizations such as the National Organization of Gay and Lesbian Scientists and Technical Professionals (NOGLSTP).

Nikki Santos-Ocampo (she/her) is a freelance creative consultant working towards the holistic development of the Philippines through the lens of creativity. As a pansexual woman and a creative that understands the struggles Philippine scientists face in developing their careers, she joined Queer Scientists PH to promote both the community and the advocacy through aesthetics that speak of the unique space the organization occupies. She also hopes to build a bridge between the creative and scientific industries to showcase the importance of intersectionality, especially in the Philippine context. Her other concurrent projects as of 2021 include urban transport group AltMobility PH, coffee farmer network the Philippine Coffee Alliance, Filipino-American collective Tagalikha, and independent short film *Ang Dragon Sa Capanganuran*. Instagram: @northwhitespecs.

Hervé Sauquet (he/him) is a Senior Research Scientist at the Royal Botanic Garden Sydney. His work focuses on the evolution of flowering plants, from their origin to the present. He is also actively involved in science communication and outreach through various initiatives, including the scientific coordination of a free Massive Online Open Course on botany that was delivered to >100,000 participants.

Mario Carlo Severo (he/him) is a postdoctoral researcher at Leiden University, the Netherlands, where he works on refining attention bias modification (ABM) training in anxiety disorders using neuromodulation and psychophysiological techniques. He earned his PhD in Psychology at Ghent University, Belgium, focusing on the influence of motivational and contextual factors on the neural correlates of performance monitoring by combining electrophysiological techniques and principal component analysis. He has also worked on his broad research interests – from gender and sexuality, social emotions, to empathy – some of which spawned published collaborations with other researchers. ORCID: Mario Carlo Severo (0000-0001-7403-819X).

Emily Standen (she/her) is a PhD candidate at the Centre for the Public Awareness of Science at the Australian National University. Her research is focussed on the intersection of culture and science communication, primarily in the South Pacific. Working with science communicators in the South Pacific, Emily is incorporating culturally diverse ways of knowing and doing to the communication of science, in order to speak back to Western science communicators. As a proud queer woman, Emily has been an inaugural co-chair of the Australian Capital Territory chapter of Queers in Science – a national network supporting queer people in STEM within Australia. ORCID: 0000-0003-3445-6260. Twitter: @mlestanden.

Carla Suciu (she/her) is a science communicator and public engagement professional based in Glasgow. A bubbly bisexual disabled woman in STEM, she prides herself on being an open advocate for LGBTQIA+ issues, women's issues (inclusive of trans women), and mental health. She currently works at the University of Edinburgh creating content for public engagement, doing freelance work such as producing Science is a Drag and in her spare time spends time with her Giant African Land Snails and tortoise.

John Noel Viana (he/him) is a postdoctoral fellow at the Australian National University and a visiting scientist at the Commonwealth Scientific and Industrial Research Organisation. He is currently exploring equity and diversity issues in precision health research. His latest publication in the *American Journal of Bioethics* highlights the need to empower and engage racial minorities in pandemic policy and planning. He is also interested in ethical and societal issues in neuroscience, genomics, bioprinting, and dementia research and care. He actively organizes events for queer people in Canberra, Australia. ORCID: John Noel Viana (0000-0002-4004-7546). Twitter: @John_Noel_Viana.

Leo P. Visser (they/them) is a Māori, non-binary Canberra-based thinker and writer with a passion for good makeup and good politics. They are a drag quing and perform under the name Feral Streak. They are an avid consumer of science-themed fiction and are active in many online queer communities. They have recently volunteered with disability organization Science Alliance. Instagram: @feralstreak.

Lee Wallace (she/her) is Associate Professor in Gender and Cultural Studies in the Faculty of Arts and Social Sciences at the University of Sydney and a former ARC Future Fellow. Her research focuses on sexuality, representation, and space. Her most recent books are *Reattachment Theory: Queer Cinema of Remarriage* (Duke 2020) and, with Scott Herring, *Long Term: Essays on Queer Commitment* (Duke 2021).

Franzisca Weder (she/her) is a Senior Lecturer at the University of Queensland, Brisbane (Australia) who researches, writes, and teaches in the area of Sustainability Communication and Corporate Social Responsibility. In her milestone publications like *The Sustainability Communication Reader: A Reflective Compendium* (2021, Springer) and a 2020 paper in *Frontiers in Communication* on individual perceptions of sustainability, Franzisca follows critical and social constructivist approaches to make sustainability 'uncomfortable', to look at communicators who talk about an 'inconvenient truth' and thus cracking the existing Western and capitalist worldview of sustainability.

Acknowledgements

For many reasons, this book has been a product of love, tears, and isolation. The editors would like to express our sincere gratitude to the contributors, to series editor Clare Wilkinson, and to the team at Bristol University Press especially Paul Stevens, for enabling and encouraging this book to happen. We also thank our partners, families, and friends, our queer and science communication communities, and the colleagues who supported us through this process.

Terminology and Sensitive Content in This Book

Dear readers.

People familiar with any writings in the LGBTIQA+ space today will be aware that preferred terminology and language are constantly evolving and can change rapidly. LGBTIQA+ people are diverse, have different preferences, and attribute different meanings to terms. For these reasons, we have chosen not to provide a glossary. Any glossary we prepared would soon be outdated and could risk shutting down the conversation rather than opening it up. We instead invite readers who are learning about this space to use the many online resources that explore the meanings of LGBTIQA+ terminology.

Some of the topics discussed in this book are sensitive. By its nature any queer-themed book is likely to discuss and reflect on different forms of queerphobia, homophobia, transphobia, biphobia, and so on. This book is no different. In some chapters, these topics are central themes because the chapters discuss about how science, medicine, and technology have perpetuated the stigmatization and oppression of LGBTIQA+ people, in some cases using violent material means such as conversion therapy. These were not always easy topics for the contributors to write about and may sometimes be difficult to read. Because of this, we are flagging the issue here. We have not added content warnings about these matters for each chapter, because the themes recur throughout the book.

However, Chapter 2 and Chapter 11 both briefly engage with sensitive topics beyond what might be expected for this book's theme of queering science communication, so we have added a content warning to each as a courtesy to readers. Chapter 2 discusses an abusive 20th century scientific experiment that involved children and also suicide. Chapter 11 deploys sexual metaphors and discusses Francis Bacon's use of a rape metaphor to promote science in the 17th century.

We hope that this book opens the way for more conversations on the value of queering science communication. Take care of yourselves. We look forward to hearing your thoughts.

Introduction

Tara Roberson and Lindy A. Orthia

It seems appropriate that when we started to write the introduction for this book about queering science communication it was Mardi Gras season in Australia. Mardi Gras is an annual celebration of queer protest, liberation, and visibility that takes place in Australia's largest city, Sydney. Like many things queer, when Mardi Gras began life in 1978 it was with a long overdue burst of passionate outrage in the guise of fun, which resulted in police violence against queer people and many arrests. Since then, Mardi Gras has become a globally renowned tourist destination that injects millions of dollars into the economy, it is broadcast live on mainstream Australian television, and it has inspired many other queer events worldwide (Markwell, 2018).

As authors Alice Motion and Hervé Sauquet describe in their contribution to this book, the Mardi Gras programme also now incorporates science-themed events, demonstrating that queer people are everywhere and science communication is relevant to everyone. Queer people face specific issues in the world of science and have unique contributions to make to science communication, so bringing 'queer' and 'science communication' together in dedicated events, products, and networks is an important part of queer protest, liberation, and visibility. We hope this book will serve those aims.

Yet an examination of the science communication research literature reveals that little has been published on the intersection between queerness and science communication over the past thirty years or more (Roberson and Orthia, 2021). This is despite the damage Western science has inflicted on queer people during its long history as self-appointed 'expert' on all things queer. Nor have queer people's protests against that damage made many academic headlines in science communication research. Even within the recent flurry of scholarly interest in equity, access, and diversity in science communication (Finlay et al, 2021), queer themes have thus far rarely rated a mention. Mardi Gras may have become mainstream, but there is still work to be done.

It is remarkable that it has taken until now for a book like this to exist. Anecdotal evidence points to a large number of lesbian, gay, bi+, trans, gender diverse, non-binary, intersex, agender, asexual, aromatic, and queer (LGBTIQA+) people working in the science communication field and indeed being drawn to it from other fields of science and technology (Roberson and Orthia, 2021). This book's contributor list is a testament to that. But as in other societal domains (McNeil, 2021), it appears that these queer folk have been rather hidden in plain sight, their needs unattended to. In 2021 we published a commentary that attempted to make a small contribution to making those needs visible (Roberson and Orthia, 2021). However, as a commentary it was of necessity limited in scope and, with only the two of us as authors, unable to reflect a greater diversity of queer perspectives. For these reasons a book on queer themes and science communication is timely, if not well overdue.

Queer as noun, queer as adjective, queer as verb

So what does 'queering science communication' mean exactly? The answer is complex because the use of the word 'queer' has evolved in recent decades and continues to do so.

A short history of 'queer'

While traditionally a pejorative term in English, a new generation of lesbian, gay, bi+, trans, and gender diverse activists in the 1980s and 1990s loudly reclaimed 'queer' for themselves across the Western-Anglophone world and beyond through networks such as Queer Nation, a spin-off of ACT UP (AIDS Coalition To Unleash Power). The same era saw the advent of the academic field of 'queer theory' (Sullivan, 2003) whose scholars sought to understand the socio-political and cultural dynamics of queer people's lives and to challenge hegemonic cis-heteronormativity in academia and beyond.

At that time, the word 'queer' took on a radical inclusivity via what was essentially a collective gesture of solidarity among many LGBTIQA+ people. 'Queer' became available for anyone in the non-cishet rainbow to use for themselves if they wanted to, and for diverse people marginalized by sexuality and gender to unite under a common queer umbrella. Some LGBTIQA+ people objected to using a word that was once a slur and some still do. However, for others of us, it was and is a useful word because, unlike Western-Anglophone alternatives like 'lesbian', 'gay', 'bisexual', or 'transgender', the word 'queer' does not imply a particular sexual orientation or relationship to gender. It comfortably embraces non-binary, agender, and pansexual identities, and other gender and sexuality variants, more than the other available terms. People whose sexuality or gender is fluid or in transition or uncertain can use the 'queer' label without worrying that they do not fit in it. People who are asexual or aromantic can also embrace the term as a signifier that their relationship to sexuality is marginalized within mainstream heteronormative society, just like that of other queer people (Scarlett, 2020).

That doesn't mean the word queer is a solution to all problems. It has a Western-Anglophone association and history that not everyone in the world will be comfortable with or interested in. In addition, even within the Anglophone West, the use of 'queer' to describe people with variations in sex characteristics, or intersex people, is inappropriate if it is done irrespective of those people's sexuality or gender (Intersex Australia, 2021). However, intersex people share common issues with LGBT people and of course many intersex individuals are also queer (Intersex Australia, 2021). Thus, while it is important to avoid subsuming intersex people under the queer banner, many of the issues intersex people face are queer *issues* – issues related to the power dynamics of binary-cis-heteronormativity. Despite its limitations, 'queer' remains a useful word for many circumstances involving LGBTIQA+ people.

The many dimensions of queering science communication

In this way, the adjectival form of 'queer' can apply not just to people but also to other things such as queer concerns. Therefore, while 'queering science communication' partly entails attending to our field's engagement with queer people, it also entails attending to science communication's engagement with queer needs, experiences, perspectives, knowledge, skills, expertise, and more. And it means thinking about how queer people, queer concerns, and queerness itself are represented in our public science communication products, from museum exhibits to tweets to movies. So our first step when queering science communication is to challenge the assumption that everything and everyone is binary-cis-hetero unless explicitly identified otherwise, a basic tenet of queer theory (Fox, 2013). Many contributions to this book are consistent with that project.

Queering science communication must go beyond mere *inclusion* though. Some scientists and communicators have built their careers studying, writing, and talking about LGBTIQA+ people, but often in an oppressive or exploitative way. Being included as an object of study is not the same as exerting control over science and its communication. This agency for LGBTIQA+ people is vital and necessary if we are to illuminate failures and flaws in current systems. For example, in a discussion of bioethics and intersex advocacy, Carpenter and Jordens (2022) confront the failure of bioethics to focus on forced and coercive medical interventions on intersex bodies because bioethics 'experts' have so far failed to understand that being intersex is not a pathology, it is a range of sex variations that have become pathologized via stigmatizing medical and social discourse. As Carpenter notes (Carpenter and Jordens, 2022, p 110), 'Intersex bodies have been medicalised as a way of dealing with what is perceived to be a social problem. The historical evidence of other ways of responding to the existence of intersex people, of including intersex people, shows that there are other ways of responding to that existence today.' Genuine inclusion often requires a fundamental shift in power dynamics of this kind.

Some queer theory adherents would suggest we go still further, to undermine the normative in general. As Gunckel (2009, p 63) put it, 'At its core, queer theory is about making the normal queer. It questions the normative processes that structure lives, actions, language, power, and knowledge.' Those normative processes are usually invisible for the people who fit within them. For example, cisgender people (people whose gender corresponds to the sex they were assigned at birth) are unlikely to notice how cisnormative structures affect the lives of trans and gender diverse people. Likewise, heterosexual people can be blind to the way heteronormative practices impact people who are homosexual, bisexual, pansexual, asexual, and so on. It is in this sense that 'queer' is used as a verb. To queer science communication is to make visible these boundaries and restrictions that can tend to go unseen, to expose their problematic impacts on LGBTIQA+ people, and ideally to root them out and nurture queerer alternatives in their place.

Finally, in an intellectual sense, 'queering' can sometimes be used metaphorically to mean dismantling existing power structures, irrespective of their relevance to LGBTIQA+ people. To queer something is to challenge the very notion of 'knowledge as mastery' and ideologies such as 'certainty, linear thinking, efficient causality [and] impermeable categories' (Bauman, 2015, p 748). Some contributors to this book have taken this challenge seriously, with an approach to their writing that continually interrogates their own positionality and attempts to radically shift authority away from its current hegemonic strongholds, including academia and science themselves.

Representing LGBTIQA+ diversity in the book

All of these considerations have meant that as editors, we also had to grapple with how the book represents queerness and whose voices it amplifies. We are two queer people working in science communication, but the queer perspectives from which we can speak remain narrow. In editing this book, we sought to avoid compounding a known issue in the LGBTIQA+ community where queer discourse can fail to adequately represent the incredible diversity of voices available, including the voices of gender minorities, intersex people, First Peoples, people of colour, and people from the Global South. It would also have been a less compelling, less joyous experience if we had not had the opportunity to work with the 43 other people who have contributed to this book. The contributions comprise an incredible array of work, drawing from different domains of science communication, different queer identities within the LGBTIQA+ rainbow, and international perspectives drawn from the world's nations and cultures. And yet there is still more work to be done and more people to hear from! We hope this is the beginning of research and discussion on queering science communication and not the end.

The LGBTIQA+ community is a diverse entity. In fact, it would be more accurate to refer to LGBTIQA+ communit*ies*, despite the idealistic ambitions of the umbrella term 'queer'. People within the LGBTIQA+ rainbow may also belong to one or multiple identities within it, which necessitates an intersectional approach to inclusion (Crenshaw, 1991). For instance, a person might be queer, non-binary, and asexual. The book addresses some of this diversity but it has limitations. While we have strived for openness and inclusion in our editorial practices, all scholarship is a product of its creators, and the book as it stands does not attend to all identities in the LGBTIQA+ rainbow equally. The book does a reasonable job of covering topics related to lesbian, gay, and bi+ identities and attractions, and topics related to gender diversity and trans and non-binary identities, because these identities are well represented among our contributors. But it gives less attention to the needs of people with intersex variations and people who are asexual or aromantic. This is another reason much more work is needed in this field. Lest we forget that science communication products are littered with binary assumptions about sex characteristics, and assumptions about human sex drives and relationships. Much of that discourse ignores the realities of intersex, asexual, and aromantic people's lives.

When deciding on key terminology for the book, we made the decision to use LGBTIQA+ as a standard term throughout unless a topic or example is focused on specific communities, in which case acronyms and terms will vary (for example, LGBT, LGBTQ+, queer, or other subsets). We made this decision to assert our stance that trans, non-binary, gender diverse, intersex, asexual, and aromatic folk are part of the queer family to whatever extent that they want to be part of it. Part of our motivation was to distance ourselves from the anti-trans tendencies that are unfortunately active among some LGB groups and which we wholeheartedly oppose. But we acknowledge that using the full acronym is sometimes just lip service when the needs of some groups (especially intersex, asexual, and aromantic folk) are not addressed. So, the other reason we use the acronym is to also remind ourselves to continually improve and make further efforts to include all LGBTIQA+ communities with respect.

The book's contributors are not just diverse in the sense of the LGBTIQA+ communities we are part of. We include people with disabilities and neurodiversities. We come from all parts of the science communication world. Our professional identities include researchers and lecturers, practitioners of outreach, science writers and journalists, drag performers, artists, activists, science museum staff, scientists, and healthcare workers. We live in multiple parts of the globe, including Austria, Australia, England, Germany, India, Ireland, Scotland, Spain, Sweden, the Netherlands, the Philippines, and the United States. Some of us proudly hail from Global South cultures and some are First Nations people living under Western colonial regimes.

We are proud of this book and the diversity of its contributors, but we remain aware that it merely scratches at the surface of the wonderful diversity that is gender, sexuality, sex characteristics, and gender expression, and the relationship of these to science communication. There is much more to do and we look forward to seeing what happens in the wake of this and other works.

Inventing a subgenre during a global pandemic

No early 2020s book would be complete without mentioning COVID-19 somewhere. It is no small thing that we – editors and contributors – entirely wrote this book amidst an ongoing, never-ending global pandemic. On top of the universally difficult and jarring experiences prompted by this worldwide crisis, some measures have made it especially painful for LGBTIQA+ people. Public health measures have harmed our communities by outing individuals in hostile environments, forcing people into dysmorphic and dangerous gender-segregated lockdowns or unsupportive households, preventing vital connections through distancing regulations, magnifying structural employment and income disparities, and magnifying healthcare disparities for people who are often already wary of seeking medical assistance in any circumstance (Burgess et al, 2021; UN Human Rights, 2021). During all this upheaval, this book has been a crucial point for continuing connection with LGBTIQA+ colleagues, friends, and extended family.

The pandemic has also had specific impacts on science communicators, whether in academic or professional practice. Our contributors, our potential contributors, and we ourselves certainly experienced painful and distressing disruptions to our lives. Because of this, not everyone could continue to contribute to the book over the course of time so their contributions to this topic will have to wait for a future opportunity. Communication was more often than not patchy during the writing and editorial process, as contributors changed jobs and lost jobs, research activities and performances were cancelled, and people in our lives became ill or passed away. In the face of this adversity, what has resulted is a testimony to the drive of our contributors to get these ideas out there because they are too important to let go of. At least some of this motivation is due to a genuine need and desire for social change in how science communication exists and operates in the world.

But COVID-19 merely added to the difficulties we already faced. The distinct queer gap in the science communication research literature that we mentioned earlier was the rationale for the book as we have stated, but it also created a problem. The usual, reliable touch points of academic writing – that is, some (!) agreed upon contexts and concepts, defined topics, or methods of investigation – were simply not out there. Consequently, we and the other contributors have had to build our topics from the ground up, literally inventing a new literature on this topic – a new sub-genre of science communication discourse. We were lucky in this process to have some contributors who had pioneered their topics in PhD and other research projects or in the practitioner arena, but in many cases contributors did not have even that 'luxury', and their contribution to this book was the first thing they have formally written on the topic.

As a result, the book is not a straightforward 'authoritative overview of the field' as might be expected for a volume in a series about Contemporary Issues in Science Communication. Until now there was no such 'field'. While we produced the book with the vision in mind that each of the 11 research chapters would be an authoritative queer perspective on a science communication domain, and we strived always for the highest quality of academic work, inevitably authors had to draw on whatever was relevant from other disciplines, such as queer theory, and from professional practice, personal experiences, and more. Looking back, we are impressed everyone managed to produce the exciting work evident in this final product, given the dynamism and fluidity of the space we were working in, with multiple meanings and languages, and much conversation and iterative writing required.

Personal experiences ended up flavouring most chapters, because queering science communication is more than an academic exercise for many of the contributing authors. A topic so inherently infused with the need for protest, liberation, and visibility cannot have the 'personal' extracted from it without losing something important. For this reason we gave contributors (including ourselves) freedom to write in first person if desired. The objectifying gaze of science has done too much damage – and queer people have been without voice for too long – for us to insist that queer authors remove themselves from the narrative that we have here begun to recount. So, while our work has rigour behind it, the tone at times departs from the dry academic style readers may expect, sometimes considerably. We hope this is a strength that makes the text more engaging to read, while also moving the field along via lateral approaches that queer academic norms.

Since science communication is a practical domain, we also wanted to ensure the book remains relevant to practitioners. As such, it includes 14 shorter 'practice spotlights' that highlight queer-themed events, activities, exhibitions, organizations, networks, media products, and short, sharp research findings of relevance to science communication professionals. They are diverse, even eclectic, but always interesting and often very inspiring. The final list of practice spotlights was determined by the contributions we received rather than any attempt to be representative of the field and we hope they will spark much further activity in this domain.

A guide to navigating this book

The longer, research-focused chapters and shorter, practice-informed spotlights are organized for the reader in four discrete parts. Each part is united by a common theme, though there are of course overlaps and synergies between them. Every part except Part II contains three chapters and three spotlights; Part II has two chapters and five spotlights. Information about the authors, including their pronouns, is available in the 'Notes on Contributors' section towards the front of the book.

At the end of each of Parts I–IV we have included questions and activities as prompts for teaching based on the part's content. These teaching notes are not just for teachers and might also assist science communication researchers and practitioners looking to reflect on their own practice.

Part I: Negotiating Queer Identities with Science, Technology, and Medicine discusses how science and technology discourse has defined and shaped queerness historically and today, and how queer people have pushed back against it as activists, academics, and artists. First up, Aritra Chatterjee in Chapter 1 addresses medical communication as a public form of science communication that readily impacts queer people. Using examples from India, their chapter examines the framing of queerness within public psychological and HIV/AIDS discourse, demonstrating just how subjective and politically inflected this discourse is and how destructive it can be for queer people. In Part I's first spotlight, Sophia Frentz continues the clinical theme with their review of how poorly recent medical research papers concerning gender have engaged with the realities, experiences, and needs of trans, intersex, and gender diverse people. Their spotlight highlights the importance of using trans, intersex, and non-binary people's expertise to improve this. V de Kauwe and Emily Standen, in Chapter 2, examine scientific discourse about queer people from the other side of the fence: queer activism. Their chapter interrogates ways queer activists have used scientific rhetoric in their political activity over the decades primarily in the US, and the destructive consequences that can arise from doing so if the nuance of queer experience is sacrificed for the sake of pushing a specific agenda.

Following this, in Chapter 3 Tara Roberson shifts the focus to technology and innovation discourse. Her chapter examines the harms and benefits digital and other technologies afford to queer folk and how discourse in the field can allow or prevent queer contributions to technological innovation. Two spotlights round out the section, both recounting artists' interpretations of queer-related science, technology, and medicine developments, and their interventions to regain queer agency within these domains. Annalaura Alifuoco, Natalie E.R. Beveridge, Yasmine Kumordzi, and Hwa Young Jung worked together on an art-meets-science project entitled 'Endosymbiotic Love Calendar 2021' that reframed symbiotic microorganisms as radically queer beings, discussed in the first of these spotlights. They literally brought this queer-science-art discourse into people's kitchens in the form of a calendar. And Helen Kaplinsky and Jessie Krish curated and produced an exhibition at Science Gallery London entitled 'GENDERS: Shaping and Breaking the Binary', discussed in Part I's final spotlight. This exhibition foregrounded queer agency through artists' reappropriation of scientific resources in their works.

Part II: Representations of Queerness in Public Science Communication turns our attention to how public science communication outputs and products frame queerness and how they represent queer people in science. In Chapter 4, Eleanor S. Armstrong and Simon J. Lock discuss approaches to queering public science institutions including science museums, science centres, zoos, and maker-spaces. Their chapter examines such queering processes from two perspectives: the institutions' public spaces ('frontstage') and their behind the scenes organization and activities ('backstage'). In a practical example of this, Josh Davis reflects on the LGBTQ+ Natural History Tour held at the National History Museum in London, UK in 2019, discussed in Part II's first spotlight. His spotlight talks through the motivations and planning for the event as well as the fun queer natural history topics included and some encouraging audience responses. Moving from exhibitions to performance, in the second practice spotlight, Òscar Aznar-Alemany reviews the Science Queers programme led by his drag queen persona, Lana Vuli. Science Queers produces in-person and online activities featuring drag and science communication, and this spotlight recounts its successes changing hearts and minds and also its struggles with continuing queerphobia. Continuing the drag theme, in the third spotlight for Part II, Carla Suciu, Brynley Pearlstone, and Sam Langford reflect on the challenges and opportunities of their Science is a Drag! event. As their spotlight shows, Science is a Drag! platforms queer people and adopts queer art forms to elevate science. In Chapter 5, Lindy A. Orthia and Leo P. Visser present a science communication analysis of how queer characters, especially scientists, are represented in science-themed fiction. Their chapter discusses the inherent ambiguity in interpreting characters' gender, sexuality, and sex characteristics, and how a queer audience's interpretations of such representations can grant unique insights that science communicators should take seriously to more fully understand how people process fiction media. Two spotlights focused on social media complete Part II. In the first, Luis Lopez and Alberto I. Roca detail the practical tactics queer scientists and communicators use on Twitter to network with one another and reach a public audience. Their spotlight identifies relevant hashtags that have gained traction and approaches that seem to gain support, offering useful guidance for queer tweeters in STEM. In the final spotlight, Ron Buckmire and Alberto I. Roca reflect on the public impact of Buckmire's early STEM queer-related blog, The Mad Professah Lectures. The blog proudly furthered queer visibility in science and also sought a broad audience for events of significance in Black gay circles, and complex issues related to HIV/AIDS.

Part III: Queer People in Science Communication Communities explores queer folks' struggles for inclusion and meaningful visibility within science communication professions, networks, and opportunities. The first chapter, Chapter 6, was co-authored by a team of science communicators hailing from the Philippines: John Noel Viaña, Mario Carlo Severo, Miguel Barretto-Garcia, Paul James Magtaan, Jason Tan Liwag, Roemel Jeusep Bueno, Christer de Silva, and Ruby Shaira Panela. The authors provide a powerful exploration of queering science communication in that nation, where the queer and colonized remain subjugated and dispossessed by social, economic, and cultural conditions. A spotlight about the Filipino advocacy and support organization Queer Scientists PH follows, written by executive board members Jason Tan Liwag, Jay S. Fidelino, Rey Audie S. Escosio, Almira B. Ocampo, and Nikki Santos-Ocampo. Queer Scientists PH aims to raise the visibility of queer Filipino scientists, and the spotlight discusses why this matters, appropriate strategies for doing so without risking members' safety, and the public impact the organization has had thus far. Shifting to Australia for the second spotlight of Part III, Alice Motion and Hervé Sauquet discuss an annual 500 Queer Scientists event held during the Sydney Gay and Lesbian Mardi Gras to celebrate queer scientists and their research. Their spotlight discusses the rarity of science-related activities within queer festivals and how this event has enabled some LGBTIQA+ people to finally find their place within Mardi Gras. In Chapter 7, Katherine Canfield considers the challenge of creating science communication communities that are radically and intentionally inclusive of LGBTIQA+ people and that attend meaningfully to intersectional queer identities. Her chapter discusses practical strategies for building inclusive networks and events, emphasizing the need to continue adapting to diverse people and needs. In Part III's final spotlight, V de Kauwe and Kai Fisher of the Australian disability organization Science Alliance describe an event called 'Rainbow Spectrums', presented by queer science communicators with disabilities.

The event's aim was to encourage an audience of queer disabled people to reclaim a sense of belonging within science and their spotlight recounts its successes from organizer and audience perspectives. Finally, in Chapter 8 Todd A. Harwell explores LGBTIQA+ people's experiences participating in citizen science projects in the US with recommendations for improving their experiences and broadening participation. His chapter highlights how few community and citizen science projects have focused on queer topics, issues, or participants in the US and elsewhere.

Part IV: Queering Institutional Science Communication Agendas presents challenges for science communication as it grows into a discipline with increasing institutionalization, in terms of practice, teaching, and research. First, in Chapter 9 Alice Motion and Lee Wallace examine the unique contributions queer people can make to science communication outreach. Their chapter highlights the important roles that storytelling, code-switching, and performance have played in queer communities, and the critical value of these skills in contemporary science communication practice. In the first spotlight for Part IV, Franzisca Weder draws on interviews she conducted with 25 queer science communicators in different countries that problematize dominant communication conventions when it comes to environmental communication. As her spotlight reveals, the communicators considered stepping on toes and asking awkward questions to be core responsibilities, turning science communication into advocacy communication. In the second spotlight, Sarah Durcan and Andrea Bandelli of Science Gallery International in Ireland reflect on how extensive LGBTIQA+ representation on the Gallery's senior management team and board influences their approach to science communication. The open, queer-positive dynamic of their workplace led to staff establishing the LGBTQ+ STEM Day that is now a worldwide initiative. In the book's final spotlight, Eleanor S. Armstrong and Sophie Gerber showcase a best practice workshop, 'Outer Edge: Queer(y)ing STEM Collections', that queered the relationship between science communication theory and practice in museums. Their spotlight offers a series of reflective questions for readers who wish to organize LGBTIQA+ inclusive events. Chapter 10's Simon J. Lock and Eleanor S. Armstrong then consider what a queer pedagogy of science communication might look like. Their chapter explores what queer theory could bring to science communication when taught as a subject in tertiary education and outside of that context. Finally, in Chapter 11 Lindy A. Orthia and V de Kauwe offer parallel serious and playful takes on what queer perspectives offer to science communication theory. Their chapter challenges the science/public binary that underlies most science communication models, using two case studies of queer people participating in scientific research design and a smutty, metaphorical exploration of science communication's heteronormative relationship with Western science.

In the book's conclusion we bring together thoughts on all the contributions and draw out some unresolved issues that deserve attention in future research and practical science communication endeavours.

References

Bauman, W.A. (2015) 'Climate weirding and queering nature: getting beyond the Anthropocene', *Religions*, 6: 742–754.

Burgess, C.M., Batchelder, A.W., Sloan, C.A., Ieong, M., and Streed Jr, C.G. (2021) 'Impact of the COVID-19 pandemic on transgender and gender diverse health care', *The Lancet*, 9(11): 729–731.

Carpenter, M. and Jordens, C.F.C. (2022) 'When bioethics fails: intersex, epistemic injustice and advocacy', in M. Walker (ed) *Interdisciplinary and Global Perspectives on Intersex*, Cham: Springer International Publishing, pp 107–124.

Crenshaw, K. (1991) 'Mapping the margins: intersectionality, identity politics, and violence against women of color', *Stanford Law Review*, 43: 1241–1299.

Finlay, S.M., Raman, S., Rasekoala, E., Mignan, V., Dawson, E., Neeley, L., and Orthia, L.A. (2021) 'From the margins to the mainstream: deconstructing science communication as a white, Western paradigm', *Journal of Science Communication*, 20(01): C02.

Fox, R. (2013) '"Homo"-work: queering academic communication and communicating queer in academia', *Text and Performance Quarterly*, 33: 58–76.

Gunckel, K.L. (2009) 'Queering science for all: probing queer theory in science education', *Journal of Curriculum Theorizing*, 25: 62–75.

Intersex Australia. (2021) 'Intersex for allies', *Intersex Human Rights Australia* [website] 7 March, Available from: https://ihra.org.au/allies/ [Accessed 6 March 2022].

Markwell, K. (2018) 'How the histories of Mardi Gras and gay tourism in Australia are intertwined', *The Conversation* [online] 2 March, Available from: https://theconversation.com/how-the-histories-of-mardi-gras-and-gay-tourism-in-australia-are-intertwined-92733 [Accessed 5 March 2022].

McNeil, P. (2021) 'Friday essay: hidden in plain sight – Australian queer men and women before gay liberation', *The Conversation* [online] 5 March, Available from: https://theconversation.com/friday-essay-hidden-in-plain-sight-australian-queer-men-and-women-before-gay-liberation-155964 [Accessed 29 February 2022].

Roberson, T. and Orthia, L.A. (2021) 'Queer world-making: a need for integrated intersectionality in science communication', *Journal of Science Communication*, 20(01): C05.

Scarlett, A.O. (2020) 'Asexuality is the queerest thing', *Stonewall* [online] 27 October, Available from: https://www.stonewall.org.uk/about-us/news/asexuality-queerest-thing [Accessed 6 March 2022].

Sullivan, N. (2003) *A Critical Introduction to Queer Theory*, New York: New York University Press.

UN Human Rights. (2021) 'An LGBT-inclusive response to COVID-19', *Office of the High Commissioner for Human Rights* [website], Available from: https://www.ohchr.org/EN/Issues/SexualOrientationGender/Pages/COVID19LGBTInclusiveResponse.aspx [Accessed 4 February 2022].

PART I

Negotiating Queer Identities with Science, Technology, and Medicine

1

Where to 'Keep' the Queer: Contestations and Anxieties in Clinical Communications

Aritra Chatterjee

Given its inevitable influence in all stages of our lives, medical discourse is a significant form of public science communication and one that is especially relevant to queer people. Queerness has often been included within dominant discourses of pathology in public medical discourse and queer people have been at the active receiving end of this institutionalized oppression. For instance, the implicit assumption of cis-heterosexuality as the by default biological/clinical norm continues to otherize and alienate queer people in various capacities in clinical communications. This of course runs hand in hand with cissexist and heterosexist societal values that are mirrored by major social institutions of medicine, law, heteronormative family structures, and religion.

In this chapter, I examine the construction of queer identities in clinical communication along two distinct disciplinary trajectories: first, the framing of queerness as mental illness by the mental health sciences; and second, the idea of 'risky sexuality' in public health discourse centring the AIDS epidemic. It is convenient and tempting to see clinical communications arising out of public medical discourse as inarguable facts that are objective, neutral, and apolitical. Yet histories of knowledge production regarding queer identities in clinical disciplinary frameworks speak otherwise. This will be my central argument in this chapter. In this context, I will be looking at the evolution of clinicians' characterizations of 'homosexuality' from a site of pathology to a normal variant of human sexuality. I will also be taking note of the radical shift in clinical communications from pathologizing of transgender identities to acknowledgement of the gender spectrum and the diversity of gender identities, expressions, and experiences. For transgender individuals, the current clinical consensus is focused on alleviating dysphoria or distress (if any) due to felt incongruence between their sex-assigned-at-birth and experienced gender.

With changing frames of clinical understandings, the reference points for what constitute the binaries of the 'clinically normative' and the 'deviant' subjects also keep shifting. This shift has massive implications for it is accompanied by rapidly changing frameworks of care. 'Conversion' of queer individuals to compulsory cis-heterosexuality was once recognized as 'cure' and now the same has been discredited by major mental health professional bodies around the world. Instead, affirmative frameworks of care continue to develop and replace such malpractices in the interest of harm-avoidance, ethical care, and respect for dignity and integrity of queer individuals. With rising evidence of AIDS as a public health concern, clinical communications target safe sex practices in general instead of holding one's gender/sexuality accountable. These developments in clinical communication attest to the truism that clinical communications are very much a product of human thought, and are open to reformulation and contestation, rather than being unshakable and absolute.

It might be misleading to think of these paradigm shifts in disciplinary understanding of queer subjects as automatic, organic, and linear. This drives home the idea that these changing frames in clinical communications came willingly from within these disciplines themselves without any external influences. Once again, histories of knowledge production regarding queer subjects bear testimony to the emptiness of such claims. They bring to light how rights-based movements, critical thinking, social justice orientation, and queer-feminist pedagogies have repeatedly maintained checks and balances over the absolutism of clinical communications, negotiating what is communicated and in whose interests. Rose (1988) notes the importance of 'social contestation' in the emergence of 'truth', in which resources and technologies in the form of results, evidence, empirical observations, and arguments are deployed in an attempt to win allies and force something into 'the true'. The spirit of social contestation embodied by queer activism is a necessary antidote to the queerphobia in dominant clinical discourses and has been instrumental in changing the discourse from the grassroots up. It is therefore fundamental that such tensions are acknowledged rather than conceptualized as a progressive refinement in disciplinary understanding.

The bulk of public clinical discourses that I shall be alluding to have been developed in Western contexts. As a result, the cultural shifts that occur in discourse produced outside the West are often overlooked and understudied and a predominant white Euro-American-centric gaze dominates conversations. In this chapter, I will locate the trajectory of these conversations in a non-Western context and the corresponding ways in which queer identities have been constructed in public science communication. By using India as a case study, I aim to identify any additional nuances in this ongoing, international conversation and decolonize the understanding of queer identities in clinical communications.

To pathologize, neutralize, or affirm: the dilemma with queerness in mental health sciences

Clinical communication in mental health sciences hangs heavy with causative explanations of homosexuality and trans identities along the lines of 'gay gene', inadequate parenting, over-identification with opposite-sex authority figure, experiences of childhood sexual abuse, and the list goes on. Contrary to many other scientific discourses, such research was never safeguarded from public scrutiny, in fact often strikingly popularized in dailies, virtual platforms, and other forms of popular media (Roberson and Orthia, 2021). In fact, these pathological explanations birthed a tradition of curative violence that made inroads into professional circles and influenced interventions (Drescher, 2015).

Homosexuality made an entry into the first edition of Diagnostic and Statistical Manual for Mental Disorders (DSM) (APA, 1952) as a 'sociopathic personality disturbance.' Sexual rights activism was already gaining grounds in the US and the Stonewall movement in 1969 added momentum to the same. Queer rights activists perceived contemporary clinical communication to be a potent contributor to social stigma and disrupted the 1970 and 1971 annual meetings of the American Psychiatric Association (APA) to make their concerns visible before the mental health fraternity (Duberman, 1994).

Historically, as psychiatrists, physicians, and psychologists tried to 'cure' homosexuality, a bulk of evidence-based research surfaced in the mid-20th century that included non-patient populations and attempted to normalize same-sex attraction (Drescher, 2015). The Kinsey reports found same-sex attraction to be more commonplace in the general population than was popularly believed, and revolutionized contemporary conservative ideas about human sexuality (Kinsey et al, 1948, 1953). Hooker used hardcore statistical and psychometric methods to refute claims of psychological maladjustment in gay men compared to heterosexual counterparts (Hooker, 1957).

At this juncture, the APA, faced with a rising tide of activism and evidence-based research, was persuaded to engage in deliberations about the status of homosexuality as a psychiatric diagnosis and its Board of Trustees voted to remove homosexuality as a mental disorder in 1973. The APA's removal of homosexuality as an illness category only goes on to underscore the bio-medical power wielded by psychiatry in its ability to classify aspects of being as illness. A raging criticism of this move came from the psychoanalytic fraternity on the grounds that the APA had caved into political and social pressure over empirical proof and scientific rigor (Greenberg, 1997). The language of this criticism only reiterates a tendency to isolate clinical communications from their social ramifications and escape accountability. By drawing an analogy between the removal of homosexuality as a psychiatric diagnosis and the 2006 removal of Pluto as a planet, Drescher (2015) turns the argument around to demonstrate the element of human subjectivity in science-making and how it filters interpretation of so-called facts.

Clinical anxieties regarding normalizing same-sex attraction continued to resurface time and again in the DSM and were legitimized in the form of diagnostic labels. The revised DSM-II printed in 1974 made reference to Sexual Orientation Disturbance (SOD) as a diagnosis reserved for those distressed with their same-sex attractions and wanting to change (APA, 1968; Spitzer, 1981). 'SOD' was rebranded 'ego-dystonic homosexuality' in 1980's DSM-III but the political compromise was quite evident (Drescher, 2015). In removing 'ego-dystonic homosexuality' from the 1987 revision, DSM-III-R, APA implicitly accepted a normal variant view of homosexuality that dominates clinical conversations of the hour (APA, 1987). A new diagnosis of 'Gender Identity Disorder in Children' (GIDC) appeared for the first time in the DSM-III (APA, 1980; Zucker and Spitzer, 2005). Multiple studies cited by Zucker and Spitzer drew the conclusion that GIDC most commonly led to a homosexual preference. Some clinicians who 'treated' children diagnosed with 'gender identity problems' cited the prevention of later homosexuality as a main treatment goal (Zijlstra, 2014). This gives more reason to wonder if the new diagnosis of 'GIDC' was a tactic to continue the stigmatization of same-sex attraction.

As far as bisexuality is concerned, most theories in mainstream psychology are rooted in an assumption that people are born gay or straight (Barker, 2007). This often contributes to bi-invisibility in public clinical communications and professionals who draw on such science communications are likely to view bisexuality as unreal, or perhaps as a phase on the way to a lesbian, straight, or gay identity (Richards and Barker, 2013).

The medicalization of aspects of existence as pathology has also been applied to transgender and gender diverse people. The diagnostic label that was initially circulated in public psychiatric discourse to target transgender individuals was 'Gender Identity Disorder' where the gender identity of the person is framed as the clinical problem in itself. The present nomenclature in the DSM-5 is Gender Dysphoria (APA, 2013), where the focus is on the affective/cognitive dissonance experienced by the trans person due to perceived incongruence between the sex-assigned-at-birth and experienced gender. The shift in terminology is important because it focuses on dysphoria as the clinical problem and spares the identity of the trans person from being pathologized. As per accepted professional standards, trans people often have to secure letters of support from mental health professionals to seek gender affirmation surgery

(Meier and Labuski, 2013). This requires a DSM-5 diagnosis of 'Gender Dysphoria', or a clinical equivalent from other classificatory systems. The debate is on about whether gender affirmation processes should at all require a medical diagnosis. However, what stands out is the titled power dynamics inherent in such protocols, where the trans person is at the mercy of the mental health professional and has no agency of their own. Further, it provides the professional enough scope to gatekeep access to affirmative services. The *International Classification of Diseases* (ICD) 11 has proposed to include 'Gender Incongruence' under sexual health and remove it from the list of mental and behavioural disorders. This perhaps points to another struggle for social contestation, whereby trans persons are not imagined as walking diagnoses, but as valid human beings with their own experiential realities.

It is important to question who writes the scripts for clinical communication in the interest of trans persons: mostly cis (also probably heterosexual) doctors. Within the psycho-medical paradigm, gender affirmation surgeries (sex reassignment as it continues to be called) were publicly communicated to be a viable option for the treatment of gender dysphoria from the mid-20th century (Benjamin, 1966). The effectiveness of any given surgical procedure was historically evaluated on the basis of how well post-operative trans people passed without being identified as transgender. However, by widely publicizing this discourse as a standalone reality, medical science also limits the possibilities of transgender embodiment (Latham, 2017). This may create an obligation to adopt a conventional script of transgender experience to have access to affirmative care, but in practice it negates the varied individualized experiences of being transgender.

This is by no means an exhaustive discussion of the dilemma with queerness in mental health sciences. Rather I have included some examples to set the tone for reflection and contemplation in this regard. On a positive note, I find it necessary to flag that the claiming of space within public medical discourse by queer individuals can lead to more participatory forms of communication. In contrast with dominant clinical narratives on sexuality, LGBT health psychology is already focused on validating and normalizing different sexualities and identifying the health, social care, and support needs of LGBT people (Alldred and Fox, 2015).

I shall now reflect on the public clinical discourse surrounding queerness in the Indian context from a mental health science perspective. This highlights the need for context-informed understanding of clinical communications in diverse geographies outside the West.

Queerness and mental health sciences in India

Public mental health discourse in India was complicit in enabling the idea of queerness as mental illness until the law forbade it. This queerphobia derives its sanctions from a highly hetero-patriarchal social context, where cis-heterosexual marriage is a dominant socialization goal and there is pervasive deep-rooted prejudice, overt discrimination, and hostility towards alternative genders/sexualities.

A few disconcerting trends can be noted in the nature of clinical representations of queerness in Indian mental health sciences that are quite tell-tale in themselves (Kottai and Ranganathan, 2019). When homosexuality was decriminalized for the first time by the Delhi High Court in 2009, literally overnight the *Indian Journal of Psychiatry* published rights-based publications including editorials. This shows just how profoundly subjective values enshrined in law can influence the clinical assertions of science. Again with re-criminalization of homosexuality by the apex court of the country in 2013, the serving President of the Indian Psychiatric Society (IPS) publicly claimed that homosexuality is unnatural. In 2018, a day before the apex court decriminalized homosexuality, the IPS released a position statement de-pathologizing homosexuality and extending support to decriminalization.

It is quite noteworthy that in its 2018 statement, the IPS did not explicitly disapprove of conversion therapies. It did so in 2020, post the death by suicide of a young bisexual woman from Kerala, a victim of such curative violence (Chatterjee, 2020). This time, the Indian Association of Clinical Psychologists (IACP), which had been absolutely silent regarding this issue so far, also joined hands with the IPS, which only highlights the hierarchies of power in clinical communication.

Further, clinical communication in India largely derives from the World Health Organization (WHO) classification of mental and behavioural disorders, *ICD-10* (1992) that still retains diagnoses such as 'Ego-dystonic sexual orientation', 'Sexual Maturation Disorder', and 'Sexual Relationship Disorder' (Narrain and Chandra, 2015). These diagnoses may in fact serve to legitimize conversion therapies for people with alternative sexualities/gender identities as already discussed. The landscape of public medical discourse in India regarding queer livings and practices is therefore undoubtedly ripe with pathologizing undertones. In a study that assessed Indian medical students' attitudes towards same-sex attraction from a sample of Year 2 and 3 students at a reputed government medical college in West Bengal, some haunting statistics come up (Kar et al, 2018). One in eight respondents reported negative attitudes; 15.9 per cent of respondents believed 'homosexuality' to be an illness, 24.8 per cent considered 'homosexuals' neurotic, 28.1 per cent considered homosexuals 'promiscuous', and 8.2 per cent thought that they posed a danger to children. This is not unforeseen as Indian textbooks continue to have dated homophobic and transphobic language even after review of the undergraduate medical curriculum by the erstwhile Medical Council of India (MCI) (Rashid, 2021). Lackadaisical attitudes of publishing bodies and dearth of adequate economic incentives ensure the circulation of such textbooks in the market. It is traumatizing for queer-trans medicos to be subjected to this in the name of medical education. Coalitions between queer/trans medical students and human-rights networks and collectives have the potential to counter such harm effectively.

Moving away from the mental health sciences, I shall now scrutinize the public health discourse surrounding sexuality that is informed by the AIDS epidemic. Meanwhile I shall also try to see whether these two disciplinary trajectories of clinical construction of queer identities can inform each other in meaningful ways.

The metaphor of 'risky sexuality' and selective audiences for AIDS-based communication

The AIDS epidemic made its presence felt in the United States in the 1980s, only a few years after homosexuality was removed as a psychiatric diagnosis by the APA. This is significant because early reported cases of the HIV virus were observed in men who were involved in same-sex sexual practices (Hymes et al, 1981). This once again created a scope for medicalization of same-sex attraction in men as a 'risky' enterprise, which often became far-fetched to make HIV/AIDS and homosexual practices synonymous with each other. Scientific research started popularizing AID or Acquired Immunodeficiency Disease as GRID or gay-related immunodeficiency disease (Altman, 1982). This also carried over to arresting and sensationalized headlines in the media such as 'Alert over "gay plague"' and '"Gay plague" may lead to blood ban on homosexuals' (Clews, 2017, p 232).

There is an interesting synergy across mental health sciences and public health in construction of queer identities: a graphic hyper focus on regulating male sexuality. AIDS communication has also been complicit in spearheading a heteronormative idea of women's sexuality, leading to the erasure of bisexual, lesbian, and queer women from discourse. When their HIV infections are discussed in research, they are perceived in a predominantly heteronormative manner. Their HIV status is therefore attributed to heterosexual intercourse or injection drug use; ironically often this heterosexual intercourse can be sexual assault perpetrated against these women out of homophobic and misogynist violence (Logie and Gibson, 2013). Although historically quite involved in the HIV/AIDS movement as advocates and caregivers for seropositive queer men, epidemiological narrative rendered queer women peripheral to the discourse, on the premise of being immune to the virus (Richardson, 2000).

The same can be said for trans people. Trans persons' unique informational needs, such as details of interactions between HIV treatment and hormone therapy, are neglected in public clinical communication pertaining to AIDS. Yet it is well-established that trans-affirmative communication materials hold promises in filling a service gap by promoting resilience and de-stigmatizing attitudes, encouraging provider engagement and transgender leadership and dissemination of information through competent media (Uhrig et al, 2019).

There is limited clinical neutrality in extension of the 'risky' metaphor to same-sex attraction as an overarching generalization. A tilted focus on male-assigned queer subjects while dealing with a public health concern and leaving out certain audiences from public clinical communication once again reiterate how the clinical is political. This is not just limited to the West, but is re-enacted in the context of India. Further, in India, AIDS-based communications seem to draw legitimacy from indigenous queer sexual practices and create indigenous subjects of targeted communications without sufficient cultural engagement. As a result, rather interestingly, new queer identities may be shaped into existence without corresponding cultural correlates.

AIDS discourse and India: takeaways

In case of India, the AIDS epidemic came to be heterosexualized through primary instances of outbreaks amongst female sex workers based in Chennai in the 1980s (Solomon et al, 2006). Section 377 of the Indian Penal Code which criminalized homosexuality was a huge barrier to AIDS activism and community-focused communication efforts until 2018 after which it was read down by the apex court (Cousins, 2019). The Human Rights Watch has highlighted the hypocritical stance of the government in emphasizing outreach to marginal communities in public statements related to HIV/AIDS, while outreach services were targeted for prosecution by law enforcement agencies (Chatterjee, 2006).

In its initial days before 2007, the National Aids Control Program focused on the risk-category of 'MSM' (men who have sex with men), which was only later dismembered into 'MSM' and 'transgender (TG)' (Datta, 2021). The National AIDS Control Programme (NACP-III; 2007–12) first made this distinction by acknowledging different prevention and care needs of transgender individuals and also embraced the heterogeneity of MSM. Assigned female-at-birth individuals do not feature in this discourse at all except for commercial sex workers. Lesbian, bisexual, queer cis-women, and trans men are erased from discourse. The nitpicking of target audiences for communication shows similar trends as in the West.

There has been an overarching tendency in South Asia-focused HIV-prevention research to indulge in a sort of lip service to the subjectivity of indigenous sexual culture. This is reflected in homogenizing the sexual script of the native queer 'males' in terms of 'culturally embedded' forms of 'active' or 'passive' sexual role in habitual anal intercourse. This archetypal script serves as a common denominator for prescriptive understandings about risk behaviour and corresponding prevention strategies. In this context, I would like to briefly touch upon the category of '*kothi*' as an indigenous subject of HIV prevention. *Kothi* is used to broadly designate gender-variant, socio-economically disadvantaged same-sex desiring males, with a usually feminine sense of self and passive role in anal sexual intercourse (Boyce, 2007). In AIDS-based communications, *kothi* is ascribed a certain matter-of-factness as a pre-existing cultural subject. However, Boyce points to ethnographic dissonances in the apparent cultural coherence of this category of self-identification that may point to overelaboration of its cultural integrity (Boyce, 2007). This raises a question as to whether there is a propensity to capitalize on intercultural subjectivity in HIV communication to produce new subjects of prevention. It seems that as a 'passive' sexual subject earmarked for vulnerability to HIV through anal sex, '*kothi ... is in some sense a creation of the [HIV-AIDS] industry itself*' (Khanna, 2009, p 49, [my emphasis]). With respect to constructions of queerness in clinical communication, this shows how new categories of queerness can be manufactured and legitimized according to convenience and the potential that clinical discourse has in deciding the fate of queer representations.

Conclusion

In this chapter, I have in essence tried to delineate the dominant anxieties that underlie the construction of queer identities in public clinical communication and make it a political enterprise. This highlights the power of public clinical discourse in determining how queerness is constructed and who gets to construct it. I have also focused on how queer rights activism and critical pedagogies have pushed the agenda for equity and radical inclusivity of queerness in popular clinical science. Interestingly, this is not just reflected in Western contexts (which tend to be read as 'global'), but are also evident when examining India as a case study. I feel that the major challenge facing science communication in this pretext is the need to own its problematic contributions to public discourse surrounding queerness and finding ways to undo the damages. It is absolutely necessary to centre queer voices and representations in clinical communications and highlight their lived experiences to transform the popular narrative from bottom-up. Communicators of public medical/clinical discourse regarding queer identities need to be mindful of defining their political standpoints clearly while communicating, rather than avoiding responsibility for communications in the name of neutrality. It is also necessary to have ethnographically sound and context-embedded constructions of queerness in public discourse in non-Western settings rather than practicing a mainstream, colonial Western gaze. Science communication must pass the microphone to queer voices to re-imagine popular constructions of queerness in public clinical discourse.

References

Alldred, P. and Fox, N. (2015) 'From "Lesbian and Gay Psychology" to a critical psychology of sexualities', in I. Parker (ed) *Handbook of Critical Psychology*, New York: Routledge, pp 200–209.

Altman, K.L. (1982) 'New homosexual disorder worries health officials', *The New York Times* [online] 11 May, Available from: https://www.nytimes.com/1982/05/11/science/new-homosexual-disorder-worries-health-officials.html [Accessed 9 July 2021].

APA. (1952) *Diagnostic and Statistical Manual of Mental Disorders* (1st edn), Washington: APA.

APA. (1968) *Diagnostic and Statistical Manual of Mental Disorders* (2nd edn), Washington: APA.

APA. (1980) *Diagnostic and Statistical Manual of Mental Disorders* (3rd edn), Washington: APA.

APA. (1987) *Diagnostic and Statistical Manual of Mental Disorders* (3rd edn Revised), Washington: APA.

APA. (2013) *Diagnostic and Statistical Manual of Mental Disorders* (5th edn), Washington: APA.

Barker, M. (2007) 'Heteronormativity and the exclusion of bisexuality in psychology', in V. Clarke and E. Peel (eds) *Out in Psychology: Lesbian, Gay, Bisexual, Trans, and Queer Perspectives*, Chichester: John Wiley and Sons, pp 86–118.

Benjamin, H. (1966) *The Transsexual Phenomenon*, Ace Publishing Company.

Boyce, P. (2007) 'Conceiving kothis: men who have sex with men in India and the cultural subject of HIV prevention', *Medical Anthropology*, 26(2): 175–203.

Chatterjee, A. (2020) 'Talking Back to Conversion Culture in Indian Psy-Circles', *Mad in Asia Pacific* [online] 24 June, Available from: https://madinasia.org/2020/06/talking-back-to-conversion-culture-in-indian-psy-circles/ [Accessed 20 July, 2021].

Chatterjee, P. (2006) 'AIDS in India: police powers and public health', *The Lancet*, 367(9513): 805–806.

Clews, C. (2017) *Gay in the 80s: From Fighting for Our Rights to Fighting for Our Lives*, Troubador Publishing Ltd.

Cousins, S. (2019) 'Legal reforms signal a change in India's HIV response', *The Lancet HIV*, 6(5): e279–e280.

Datta, S. (2021) 'India's healthcare system is still failing queer-trans people. A public health movement can change that', *The Swaddle* [online] 26 June, Available from: https://theswaddle.com/indias-healthcare-system-is-still-failing-queer-trans-people-a-public-health-movement-can-change-that/ [Accessed 15 July, 2021].

Drescher, J. (2015) 'Out of DSM: depathologizing homosexuality', *Behavioral Sciences*, 5(4): 565–575.

Duberman, M. (1994) *Stonewall*, New York: Plume.

Greenberg, G. (1997) 'Right answers, wrong reasons: revisiting the deletion of homosexuality from the DSM', *Review of General Psychology*, 1(3): 256–270.

Hooker, E. (1957) 'The adjustment of the male overt homosexual', *Journal of Projective Techniques*, 21(1): 18–31.

Hymes, K.B., Cheung, T., Greene, J.B., Prose, N.S., Marcus, A., Ballard, H., William, D.C., and Laubenstein, L.J. (1981) 'Kaposi's sarcoma in homosexual men – a report of eight cases', *The Lancet*, 318(8247): 598–600.

Kar, A., Mukherjee, S., Ventriglio, A., and Bhugra, D. (2018) 'Attitude of Indian medical students towards homosexuality', *East Asian Archives of Psychiatry*, 28(2): 59–63.

Khanna, A. (2009) 'Taming of the shrewd *meyeli chhele*: a political economy of development's sexual subject', *Development*, 52(1): 43–51.

Kinsey, A.C., Pomeroy, W.B., and Martin, C.E. (1948) *Sexual Behavior in the Human Male*, Philadelphia: W.B. Saunders.

Kinsey, A.C., Pomeroy, W.B., Martin, C.E., and Gebhard, P.H. (1953) *Sexual Behavior in the Human Female*, Philadelphia: W.B. Saunders.

Kottai, S.R. and Ranganathan, S. (2019) 'Fractured narratives of PSY disciplines and the LGBTQIA+ rights movement in India: a critical examination', *Indian Journal of Medical Ethics*, 4(2): 100–110.

Latham, J.R. (2017) '(Re)making sex: a praxiography of the gender clinic', *Feminist Theory*, 18(2): 177–204.

Logie, C.H. and Gibson, M.F. (2013) 'A mark that is no mark? Queer women and violence in HIV discourse', *Culture, Health & Sexuality*, 15(1): 29–43.

Meier, S.C. and Labuski, C.M. (2013) 'The demographics of the transgender population', in A.K. Baumle (ed) *International Handbook on the Demography of Sexuality*, Dordrecht: Springer, pp 289–327.

Narrain, A. and Chandra, V. (2015) *Nothing to Fix: Medicalisation of Sexual Orientation and Gender Identity*, India: Sage Publications.

Rashid, A. (2021) 'Do India's medical textbooks have homophobic language?', *The Wire Science* [online] 12 February, Available from: https://science.thewire.in/health/india-forensic-medicine-toxicology-textbooks-homophobic-language-queer-trans-issues-mci-nmc/ [Accessed 20 July, 2021].

Richards, C. and Barker, M. (2013) *Sexuality and Gender for Mental Health Professionals: A Practical Guide*, London: Sage.

Richardson, D. (2000) 'The social construction of immunity: HIV risk perception and prevention among lesbians and bisexual women', *Culture, Health & Sexuality*, 2(1): 33–49.

Roberson, T. and Orthia, L.A. (2021) 'Queer world-making: a need for integrated intersectionality in science communication', *Journal of Science Communication*, 20(01): C05.

Rose, N. (1998) *Inventing Our Selves: Psychology, Power, and Personhood*, Cambridge: Cambridge University Press.

Solomon, S., Solomon, S.S., and Ganesh, A.K. (2006) 'AIDS in India', *Postgraduate Medical Journal*, 82(971): 545–547.

Spitzer, R.L. (1981) 'The diagnostic status of homosexuality in DSM-III: a reformulation of the issues', *The American Journal of Psychiatry*, 138(2): 210–215.

Uhrig, J.D., Stryker, J.E., Bresee, S., Gard Read, J., Parvanta, S., Ruiz, F., and DeLuca, N. (2019) 'HIV information needs of transgender people and their healthcare providers', *AIDS Care*, 31(3): 357–363.

World Health Organization. (1992) *The ICD-10 Classification of Mental and Behavioural disorders: Clinical Descriptions and Diagnostic Guidelines*, Geneva: World Health Organization Press.

Zijlstra, I. (2014) 'The turbulent evolution of homosexuality: from mental illness to sexual preference', *Social Cosmos*, 5(1): 29–35.

Zucker, K.J. and Spitzer, R.L. (2005) 'Was the gender identity disorder of childhood diagnosis introduced into DSM-III as a backdoor maneuver to replace homosexuality? A historical note', *Journal of Sex & Marital Therapy*, 31(1): 31–42.

Practice Spotlight: Gender and Sex in Research Communications

Sophia Frentz

Medical research largely fails to consider or even acknowledge the existence of transgender and intersex people. This results in a lower likelihood of people, firstly, receiving appropriate medical care as the information to provide that care may not exist and, secondly, accessing medical care because they fear harm resulting from inappropriate medical treatments.

Transgender people already often avoid accessing healthcare due to discrimination (Jaffee et al, 2016), further compounding the worse health outcomes we experience (Cicero et al, 2020). This is particularly clear when considering procedures that may be dysmorphic, for instance gender minorities access cervical smears at a significantly lower rate than cis women (Connolly et al, 2020).

Intersex children often experience significant and ongoing medical harm. The erasure of the normal human diversity of sex characteristics in research, writing, and education entrenches stigma against the intersex community and contributes to this medical harm (Carpenter, 2018). With intersex people estimated at 0.4–1.7 per cent of the population (Carpenter, 2016), this is roughly equivalent to erasing the existence of redheads.

Scientific research necessitates precise use of language when communicating work. Using words relating to gender and sex loosely or interchangeably is done with an assumption of shared understanding for words such as 'women'.

However, this shared understanding does not exist, and the assumption falls apart when applying a critical eye. 'Women' may refer to people who: are assigned female at birth, identify as women, are 46XX, can get pregnant, can menstruate, have a vagina, uterus, and ovaries, have particular hormone levels, or any combination of the previous elements. Whether research is considering the social, medical, or biochemical elements of gender, a clear definition of the terms and limitations of work is critical for meaningful findings.

I reviewed 1,841 papers published in scientific journals in June 2019 that referenced sex, gender, or sex characteristics in humans, and 24 per cent of them used the words 'sex' and 'gender' interchangeably, while less than 5 per cent of papers acknowledged the existence of trans, non-binary, or intersex people. Even within papers about the LGBTIQA+ community, 52 per cent did not acknowledge the existence of transgender people, and 90 per cent did not recognize the existence of intersex people.

When it comes to determining the gender or sex of the study subjects, 34 per cent of papers did not describe their approach at all. Self-identification of gender was used in 23 per cent of papers, however, only seven per cent of those explicitly allowed diverse genders. Finally, while 24 per cent of papers used medical records, only one per cent of those acknowledged the existence of transgender people, and none considered the impact of requirements for the altering of gender in medical records.

Science will only improve when the scientific community as a whole engages critically with sex, gender, and queerness as a whole. This is work that doesn't just support the LGBTIQA+ community, but also improves the quality of published research.

Fixing this problem means working with trans and intersex communities. Consult with the LGBTIQA+ community as you would consult another specialist or stakeholder. Read, consider, and cite our work beyond this book and, importantly, trust us to report on our own experiences.

Guidance for defining sex in medical research is provided by a range of intersex advocacy groups, including Intersex Human Rights Australia and InterACT (US). Networks such as Queers in Science (Australia), and the International Society of Non-binary Scientists can also provide support and guidance.

Simplifying gender and sex into binaries can make research easier. However, scientific research is not the pursuit of what is *easy*. It is an endeavour in pursuit of knowledge, and that will always be incomplete until we include all humanity in our work.

References

Carpenter, M. (2016) 'The human rights of intersex people: addressing harmful practices and rhetoric of change', *Reproductive Health Matters*, 24(47): 74–84.

Carpenter, M. (2018) 'Intersex variations, human rights, and the international classification of diseases', *Health and Human Rights*, 20(2): 205–214.

Cicero, E.C., Reisner, S.L., Merwin, E.I., Humphreys, J.C., and Silva, S.G. (2020) 'The health status of transgender and gender nonbinary adults in the United States', *PLoS One,* 15(2): e0228765.

Connolly, D., Hughes, X., and Berner, A. (2020) 'Barriers and facilitators to cervical cancer screening among transgender men and non-binary people with a cervix: a systematic narrative review', *Preventive Medicine*, 135: 106071.

Jaffee, K.D., Shires, D.A., and Stroumsa, D. (2016) 'Discrimination and delayed health care among transgender women and men', *Medical Care*, 54(11): 1010–1016.

2

The Question of Queer Complexity: Science Communication and Queer Activism

V de Kauwe and Emily Standen

Content warning: this chapter discusses an abusive 20th century scientific experiment that involved children and also suicide.

Political activism is an important domain in which science communication is used to support, or oppose, social change. In the recent past, queer activism has employed science-related arguments with varying results. Like many of the topics covered in this book, the academic literature on science communication in queer activism is sparse to non-existent. Accordingly, this chapter offers a retrospective examination of some of the science communication used by queer activists in the West, identified by examining historical newspaper articles, the archives of independent queer newspapers, audio-visual compilations of queer activism curated by libraries and museums, and the extant relevant academic publications. It is hoped that this foundational research will provide a starting point for future researchers to refine and fill the gaps. A further aim is to support and improve queer activism in the future by scrutinizing the role of science communication in queer activism in the past.

Nearly 75 years ago, in 1952, the first Diagnostic and Statistical Manual (DSM-I) of the American Psychiatric Association (APA) declared anything that deviated from cis-heteronormativity was a form of pathological psychosis (APA, 1952). This included a variety of non-heterosexual attractions and activities, as explained in more detail later. The inclusion of non-cis-heteronormality as a psychotic illness persisted across different editions, through direct and indirect references until at least 1987. Activism helped direct a long process of change until the APA board of trustees elected to remove the definition of homosexuality as a form of psychosis from the DSM (Drescher and Merlino, 2007). In 1990, the World Health Organization followed suit by removing homosexuality from the International Classification of Diseases (ICD-10). It took even longer for transgender-related classifications to be removed or amended in the DSM and the ICD, again under the influence of activist pressure. In 2013 the DSM-5 dropped 'gender identity disorder' as a diagnosis but replaced it with 'gender dysphoria', which many trans activists still consider problematic (Whalen, n.d.), to the point that the Danish government officially removed the 'mentally ill' classification of trans people in 2017, overriding the DSM's classification in Denmark (Russo, 2017). Similarly, the ICD descriptors of trans experiences were changed in its 11th edition in 2018, with trans classifications moved from its 'Mental and behavioural disorders' section into a section on 'Conditions related to sexual health' (Fitzsimons, 2018).

At the time of DSM-I and post this period, queer activists in Western countries developed a suite of science-based rhetoric to advocate for LGBTIQA+ matters. This chapter will examine three such approaches and consider their political consequences. Specifically, after a brief introduction into the divisive tendencies of science communication in queer activism since the mid-20th century, this chapter focuses on the role of science communication in the following well known approaches in queer activism: the argument of free choice for consenting adults (referring to the sexual activities of non-heterosexuals), the argument that gender is a social concept, and the biodeterminist argument that queer people are 'born this way'. Although this chapter focuses heavily on events in the US, these events remain relevant to LGBTIQA+ history and advocacy internationally.

Shattering the queer community

Some have argued that queer activism, in some form or another, has existed throughout history and across the globe (Koh, 2012). This naturally goes beyond the scope of DSM-I. However, this chapter uses DSM-I as a starting point since the science communication rhetoric of DSM-I had such an enduring negative impact on the queer community.

While the language used in DSM-I and DSM-II focused heavily on 'homosexual' and 'male homosexuality', its contents pathologized and conflated broad aspects of non-heteronormative life (APA, 1952, 1968). This was even to the point of listing particular movements and fantasies; that is, some sexual positions were considered indicative of psychopathy, as was role playing, the use of sex toys or props, wearing leather, and the desire of same-sex attraction even if it was not acted upon (Gerson, 2015). Essentially, what DSM-I created was a 'hierarchy of sexual deviancies' (Uyeda, 2021). In contrast, the language used for cis-heterosexuality included 'the norm', 'natural', and 'healthy' (APA, 1952). In this way DSM-I did more than claim the deviance and psychopathy of non-cis-heterosexuality. It also officially codified the assumption that conservative heterosexuality was the norm: that is, cis-heterosexuality that didn't stray far from the strict definition of coitus (Giami, 2015). This is what the APA and DSM of the time held up as 'heteronormality'. In doing so DSM-I created a dichotomy between the APA's 'heteronormality' and the deviant other, which *seemed* backed by medical science. As the academic-activist Jack Drescher points out, the hierarchical definitions that divided non-'heteronormativity' from the rest of society were 'etiological' myths based on the 'moral underpinnings' of conservative Americans, in the guise of 'psychiatric diagnosis' (Drescher, 2010, p 431).

Explicitly, DSM-I used the term 'homosexuality' to indicate any 'psychopathic personality with pathologic sexuality' that deviated from the 'norm' through the following behaviours: 'homosexuality, transvestism, pedophilia, fetishism, and sexual sadism (including rape, sexual assault, mutilation)' (APA, 1952, p 39). As Drescher (2010) indicates, the definition itself conflates 'homosexuality' with sexual crimes, and thereby conceptually constructs non-cis-heterosexuality as deviant.

It should also be noted that, while the DSM definition made no distinction between sexual crimes and the interactions of consenting adults, neither did it recognize the diversity of individuals who fell under the umbrella term 'homosexual'. Nor did it recognize the difference between sexuality and gender variance. This left room for other publications to be more explicit in their hierarchy of sexual deviances. For example, in 1965, the World Health Organization's International Classification of Diseases (ICD-8a) portrayed homosexuality as of greatest concern and danger to others, since 'Homosexuality (male and female) constituted an extended specific category which included pedophilia, sodomy, [and] exhibitionism' (Giami, 2015, p 1131). After this was the category of 'transvestism and fetishism' (Giami, 2015, p 1131), which included any deviation from cisgender identity; this group were supposedly a danger to themselves. Next came the categories of lesser concern such as narcissism, sadism, and voyeurism, which included any form of kink sex play (World Health Organization, 1965). The major concern for this group was its disregard for 'heterosexual conjugal monogamy' (Giami, 2015, p 1131). The final category was that of impotence and frigidity, which included any form of asexuality. These were classified as somatic disorders rather than psychopathies (World Health Organization, 1965).

Soon after in 1968, DSM-II further extended the list of diagnosable sexual deviations (APA, 1968). Speculation remains over how much the DSM-I and -II, and ICD-8 influenced each other (Giami, 2015). However, it seems evident that their joint focus on categorizing and pathologizing sexual and gender diversity had an ongoing effect that continues to the present day (Gerson, 2015). Such a taxonomy of 'perversions' gained the title 'paraphilias' (Giami, 2015), with some psychiatrists categorizing around 550 distinct paraphilias (Aggrawal, 2008). Berger (2005) provides an online sample list of paraphilias such as those recognized by the DSM.

Others agree with Drescher, claiming that overly scrutinizing and ranking the deviancies of everything non-cis-heteronormative necessarily leaves heterosexuality as the unblemished ideal (Uyeda, 2021). Internationally, scholars have demonstrated how the myth of an idealized heteronormality continues to be created and reified in Western cultures beyond the US since DSM-I (McEwan, 2016).

More implicitly, the hierarchical division of 'deviances' seems to have created a splintering effect. Non-heteronormality was shattered into myriad distinctly diagnosed deviances, thereby artificially dividing the queer community (Drescher, 2010). Unfortunately, some activist responses to this seemed to have perpetuated this division. Activist Judi Chamberlain recalls how, during the 1970s and early 1980s, even groups that gathered with 'the basic premise that oppressive psychiatry must be destroyed' would often 'dissolve into endless wrangling' (Chamberlain, 1982, p 4). Decisions as to what actions activists should take became 'agonizingly slow', due to deliberations over whose voice was to be heard and for how long (Chamberlain, 1982, p 4). Chamberlain's description of fractious voices clamouring to be heard was more than the usual arguing that can happen in any group. Rather, what Chamberlain seems to be reporting was the desperation of various groups within the larger queer community all trying to establish legitimate identity and self-determination, as confirmed by an anonymous reporter who attended many such rallies (Anon., 1974). This seems at least partially linked to the artificial distinctions made by DSM-I and -II. The journalist-advocate Christianson openly linked division within the gay community of the time to the hierarchy of classifications listed in DSM-I and the assumption of its scientific credibility (Christianson, 1974a). Independent newspaper *The Gay Liberator* reported on this phenomenon: '[A]mazingly, there was little objection to this practice [of separatism], with most men apparently unaware that coalitions need not require a denial of anyone's democratic rights' (Anon., 1974, p 2). That is, solidarity amongst different groups within the queer community was endangered by the notion that the self-determination of one group entailed disenfranchising others (Anon., 1974).

This fragmentation of queer activism was serious enough for several queer newspapers to make a stand against it (Uyeda, 2021). For example, the *Big*

Mama Rag told the stories of queer people of all persuasions who had suffered due to the DSM's diagnoses. These people had all suffered '[a]trocities in institutions, in "community programs", in "sheltered workshops"' occurring under the misnomer of medical science (Chamberlain, 1982, p 4). No difference in diagnostic definition could save anyone in any queer community, for anything deemed a deviation from cis-heteronormality by medical professionals could be 'defined as "mental illness" and forcibly treated' (Chamberlain, 1982, p 4). These forced treatments included electroshock therapy, chemical castration, forced drugging, and lobotomies (Hollis, 1975). As depicted in an exhibition representing stories of coming out in Australia during the 1970s, all members of all queer communities were vulnerable to 'rejection, discrimination, abuse or even death' (State Library of NSW, 2022, p 2). Reporting on this heightened vulnerability was aimed at producing an urgent sense of solidarity in diversity.

Activist newspapers such as *The Gay Liberator* acknowledged that protests against the APA often resulted in counterproductive 'divisions between men and women, radicals and liberals, lobbyists and activists' (Compton, 1974b, p 2). So they (and other similar publications) sought to defuse the tension by questioning the assumed scientific credibility of the DSM, and the forced treatments that followed. For example, details of electroshock therapy and the side effects of the drugs used to 'treat' non-heteronormality were described in great scientific detail, evoking both an abhorrence for the treatments, as well as empathy for those thus treated (Christianson, 1974a; Compton, 1974a). Others uncovered 'quasi-science research', such as the claim that non-heteronormality was a 'congenital defect' that could be identified by 'a single pernicious gene' (Christianson, 1974b, p 8).

Kameny, Gittings, and the argument for free choice

A different debunking tactic came from the well-known queer activist Frank Kameny soon after DSM-I was published (Crain, 2020). Kameny was an established scientist with a high security clearance for work within the US military, and was also known for his strict moral integrity (Crain, 2020). When his sexual activities were discovered, Kameny was dismissed from the military and denounced as a danger to national security because 'homosexuals were a security risk with "no place in the United States government"' (Crain, 2020, p 4). Although this left him homeless and close to starvation, Kameny redirected his work ethic to refuting science-based myths about non-cis-heteronormality (Baughey-Gill, 2011). As a scientist himself, Kameny placed the burden of proof back onto the medical community. If the APA were to condemn non-cis-heteronormality on a scientific basis, then they first had to provide scientific proof that non-cis-heteronormality was indeed a disease, genetic or otherwise (Baughey-Gill, 2011).

Kameny's appeal to the value of scientific proof seems to have gained him respect with the APA and he was one of only two people asked to represent the queer community at the 1972 APA convention (Baughey-Gill, 2011). The other was Barbara Gittings, now known as the 'mother of queer liberation' (The Legacy Project, 2021a). Gittings showed the same intellectual prowess and personal moral fortitude as Kameny, but seemed more willing to put herself in danger at picket lines and rallies; she also made close connections to both activists and the people they fought for (Jennings, 2011). Gittings willingly acknowledged her sexual preferences, but like Kameny, she would not accept claims of psychopathy as the latter had no empirical proof (The Legacy Project, 2021a).

Together Kameny and Gittings were a model of solidarity in diversity, attempting to uphold the rights of as much of the queer community as possible and counteract the obvious underrepresentation at the 1972 convention (Uyeda, 2021). However, it was Gittings who realized that there was also a form of misrepresentation that needed addressing. Gittings pointed out that the convention had speakers who were medical professionals, herself and Kameny who were queer activists, 'but it did not have a speaker that was both' (Baughey-Gill, 2011, p 12). Gittings seemed to realize that, statistically speaking, it was probable that there were members of the APA who were not heterosexual and therefore sought a homosexual psychiatrist who would speak on behalf of the queer community rather than silently comply with the rest of the APA. Nevertheless, she and Kameny found it nearly impossible to find a psychiatrist who would risk the backlash of revealing their own sexuality (Drescher and Merlino, 2007). Reporter Kevin Jennings recalls that it was Gittings who finally convinced a member of the APA to acknowledge his true sexuality, and to speak as someone who was *both* a queer activist and medical professional (Jennings, 2011). However, for this medical professional to speak, he wore a full-face mask and wig, attached a voice-distorting microphone inside the mask, and wore a frilled shirt with an oversized jacket and bow to distort his body shape (Baughey-Gill, 2011; Uyeda, 2021). Such was his personal and professional fear. This man called himself Dr Henry Anonymous, but was later revealed to be Dr John Fryer (Uyeda, 2021).

With regard to the science communication presented by Gittings, Kameny, and Fryer, Jennings recalls that the impact went beyond the results of that first formal meeting of APA professionals and queer activists (Jennings, 2011). As far as the results of the conference went, the request of Gittings, Kameny, and Fryer to have 'homosexuality' (and all the sexual and gender diversity contained within this umbrella term) removed from the DSM's list of psychopathic illness, was again denied (Drescher and Merlino, 2007). Yet this was far from a defeat. Gittings and Kameny had been recognized as valid contributors and their arguments seemed to have been heard and respected. Members of the APA soon investigated their questions about the DSM's scientific credibility with regard to sexual 'deviances' and found it lacked scientific evidence (Spitzer and Fleiss, 1974). As a result, to the present day, the diagnostic reliability of the DSM in all its updated editions continues to undergo intense academic and professional scrutiny (Chmielewski et al, 2015).

In a word, DSM-I was found to be unreliable and diagnostically invalid. This allowed other members of the APA to realize that the hierarchy of sexual deviances were in fact biased value judgements with little connection to either science or logic (Johnson and Rand Herron, 1973). Moreover, there were no measures taken to minimize this bias. As Kameny had discussed, there was a need for non-heterosexual participants to have an empowered role in APA research, to curtail prejudices and promote mutual understanding (Baughey-Gill, 2011; Drescher and Merlino, 2007). Further to this point, the APA's use of 'homosexual' as an umbrella term seems to have occluded the realization that representatives of other sexualities and gender identities should also have been given a voice.

Nevertheless, the positive impact of Gittings, Kameny, and Fryer was undeniable, and was heightened by the delivery of their message, as well as its content. Science communicators and activists alike recognize that the way a point is made is as important as the message itself (Canfield et al, 2020; Fish et al, 2018). Gittings' and Kameny's arguments appealed to the APA's value in science. Their personal presentation and deportment seems to have played an equally important role. Photographs and television footage from the time show them to be calm, reasonable, yet firm and matter-of-fact (Jennings, 2011; Uyeda, 2021). Their natural personalities allowed them to present themselves in a manner that gelled with the notion of ideal scientists. In contrast, next to them was the visual horror of Fryer's disguise. Worse than the disguise itself was the understanding that it was necessary: at the time, the human rights of non-cis-heterosexuals 'were entirely obstructed by the DSM diagnoses' (Drescher, 2010, p 454). And for the APA this realization was even more damning: they had done this. Now here was one of their own members, having to use this hideous disguise to speak at his own conference. In fact, it was Gittings and Kameny who had presented themselves unexpectedly, by using their true names and not hiding their appearance (Jennings, 2011).

Free choice for consenting adults vs sin and crime

Only a year later in 1973, the APA began the process of removing homosexuality from the DSM. In reality, the 1973 change was more nominal than a whole-hearted self-correction by the APA. That is to say, the APA decided not to use the nomenclature of 'psychotic illness' with regard to the umbrella term 'homosexuality' (Johnson and Rand Herron, 1973). While many rejoiced, Gittings pointed out that this was only the beginning of the battle; the change was sudden and too much was left un-investigated (Baughey-Gill, 2011).

For example, if the diagnostic hierarchy of DSM-I was indeed a reflection of conservative value judgements, what should replace them? During the 1973 nomenclature discussions, it was decided that genuine psychiatric disorders 'must either regularly cause subjective distress or regularly be associated with some generalized impairment in social effectiveness or functioning. Clearly homosexuality per se does not meet these requirements' (Johnson and Rand Herron, 1973, p 109). Kameny and Gittings took this point to its logical conclusion: allowing a person to live in a way that fulfilled their personal identity and personal sexuality, as long as it didn't harm others, was morally good (The Legacy Project, 2021a, 2021b). This claim was founded on the research of Freud, who concluded, 'Homosexuality is assuredly no advantage but it is nothing to be ashamed of. No vice, no degradation. It cannot be classified as an illness. We consider it to be a variation of the sexual function' (Johnson and Rand Herron, 1973, p 109). Kameny and Gittings also enabled greater diversity to be addressed beyond the single umbrella term 'homosexuality', with Kameny saying, 'No matter how a particular sexual adaptation is arrived at in an adult, sexual behaviour between consenting adults is a private matter' (Johnson and Rand Herron, 1973, p 109).

Our first case study considers this rhetoric of 'free choice for consenting adults' and its implications for LGBTIQA+ people. The catch phrase 'free choice for consenting adults' was coined amidst the efforts of queer activists (State Library of NSW, 2022). It was derived from the debates and battles described in the preceding sections, together with countless lesser-known struggles.

Yet there were unforeseen side effects that continued to isolate and damage queer individuals. Gittings was right; after decades of struggle, her and Kameny's interactions at the 1972 APA conference had brought a suspiciously sudden recantation at the 1973 conference – activists should not yet drop their guard. Decriminalization still had to be fought for, in some countries for decades to come (Meyer-Bahlburg, 2010).

Most significantly, the rhetoric of consensual choice inadvertently complemented the rhetoric used by conservative religious groups to characterize being queer as sinful, as it implied an active choice (Weber, 2021). The language conservative Christians used to describe this sinfulness was often potent enough to sway even the non-religious; particularly the condemnation of non-heteronormality as an abomination (Sklar, 2018). In turn, this quasi-religious attitude fuelled existing prejudices in the wider population, and queer people continued to be vulnerable to discrimination and harm from practices such as conversion therapy (UN Human Rights, 2020).

The disparity between the APA's 1973 recantation, and the prejudices still embedded within the pervasive culture were, and in some cases continue to be, gaping. In these circumstances, the rhetoric of 'free choice for consenting adults' was still oppressive to many individuals. As Drescher (2010, p 441) notes, the lack of support for queer individuals caused many of them to cling to the medical model of psychopathic diagnosis and treatment: '[T]he stigma of psychiatric illness and the paternalism of medical practitioners notwithstanding, many "homosexuals" accepted, if not embraced, the medical model as an alternative to religious and legal condemnation. While some saw in the illness model hopes for a "cure".'

In this context, it is possible to see what was lacking in the rhetoric of free choice for consenting adults. It was not that Gittings and Kameny had not presented sound science communication. Rather, the socio-political context into which they delivered their arguments was not supportive of queer individuals. On the contrary, the prevailing culture at the time was so hostile that countless individuals found it impossible to make free consensual choices regarding sexuality (Drescher, 2010; Koh, 2012). Queer people were not yet truly free.

Discrimination and abuse continued to be rife, including abuses based on misuse of science communication. Members of the APA who were part of the 1973 recantation continued to abuse medical science to destroy the lives of queer individuals, by simply changing the focus of their science communication. For example, Dr Irving Bieber openly agreed with the APA recantation that 'homosexuality is not a mental illness' since psychiatric illness 'must either regularly cause subjective distress or regularly be associated with some generalized impairment in social effectiveness or functioning' (Johnson and Rand Herron, 1973, p 9). At the same time, however, Bieber argued that not conforming to heteronormality did indeed cause a high level of distress, together with an inability to function within the prevailing culture of discrimination (Johnson and Rand Herron, 1973). Therefore Bieber continued to offer medical 'treatments' and even 'cures' for non-conformity to cis-heteronormative standards (Baughey-Gill, 2011; Drescher, 2010). As listed earlier, these treatments ranged from conversion therapy to mind and body altering drugs, and in last case scenarios, lobotomies. After the 1973 recantation, Bieber openly claimed a 100 per cent conversion success rate in the *New York Times*: '[W]e are psychiatrists. I am a scientist primarily ... I can pick out the entire population at risk in male homosexuality at the age of five, six, seven, eight. If these children are treated, and their parents are treated, they will not become homosexuals' (quoted by Johnson and Rand Herron, 1973, p 9). Bieber maintained this rhetoric in his academic publishing and, together with others, continued 'curing' both adults and children into the 21st century (Marmor et al, 1999; Zucker and Spitzer, 2005).

It seems clear that Bieber and the like were more concerned with defending their own rhetoric than the welfare of any member of the queer community. Here we have a vivid example of how science communication can be weaponized to serve prejudices.

Gender as a social construct

Our second case study, and perhaps one of the most devastating and overlooked misuses of science communication rhetoric, is the case of Dr John Money. This second case shifts the focus from sexual activity to gender identity. As shown in the first part of this chapter, the APA often conflated gender with sexuality, or ignored the complexities of gender identity altogether (Drescher, 2010). In contrast, Dr Money was a psychologist who claimed to specialize in the complexities of gender identity, and even presented himself as an activist in this area (Ehrhardt, 2007). In the 1960s, Money claimed to be researching the then burgeoning idea that gender was a social construct (Interact Advocates for Intersex Youth, n.d.). He did this by experimenting with the forced manipulation of the gender of identical twins, ultimately resulting in both twins suffering from severe mental illness and eventual suicide (Gaetano, 2017). During the experiments, Money convinced the parents of twin boys that gender reassignment surgery was essential for one of them and performed this surgery soon after birth. Throughout their childhood and youth, Money provided psychological 'training' for the twins and their parents, as well as hormone therapy for the child who had the reassignment surgery (Walker, 2004). It was later revealed that this even included sexual abuse conducted under recorded 'scientific' observation, to ensure that the twins would grow up to use their genitals in the way Money had determined for them (Gaetano, 2017).

Money used science communication rhetoric to declare his experiments a success, claiming to have scientifically proven that gender is a social construct (Ehrhardt, 2007). In the late 1990s Money's methods were openly criticized by both the survivors of his experiments and other researchers, particularly Dr Milton Diamond. However, Money publicly dismissed these criticisms as 'anti-feminist and anti-trans bias' (Gaetano, 2017, p 1). At the time of writing this chapter, Dr Money's arguments and experiments are still used by those who defend forced surgical interventions on intersex infants – sometimes termed intersex genital mutilation – despite United Nations statements against this practice (Carpenter, 2019). Such a stark case demonstrates that even science communication around a seemingly empowering concept – gender being a social construct – can be deployed in such a way as to disenfranchise the most vulnerable amongst us. In the case of Dr Money, pseudo-science was used to destroy the lives of the twins who were experimented on and misuse of science communication along these lines continues to destroy the lives of intersex children today (Walker, 2004).

While the argument for 'free choice amongst consenting adults' was developed by those who had the interest of the queer community at heart, Dr Money's intentions seem completely self-serving. In both cases, individuals within the queer community were left marginalized and in genuine danger. This helps to explain why arguments of biological determinism (or biodeterminism) became popular with some queer activists.

Born this way

In our third case study we consider rhetoric arising from biodeterminism or the 'born this way' argument used by some queer activists. Biodeterminism initially focused on the experiences of homosexual individuals and differed from both free choice and social constructivism by arguing that homosexuality was either a result of 'nature' (for example, genetics) or 'nurture' during early childhood which was then enduring and immutable for the remainder of an individual's life (Bernstein, 2002). It removed the implied 'active choice' argument that had been weaponized against the homosexual community shortly after the 1973 changes to the DSM and pushed against a solely social constructivist perspective by including the possibility of biological 'causes'. As it was immutable, and determined early in life, homosexual people had no choice or control and parallels were drawn to the inability to choose one's race (Bernstein, 2002). These parallels were seen by many American activists in the early 1960s to be a strong argument for equal rights before the law and were widely adopted.

While a biodeterminist explanation of homosexuality may appear, at first, to be highly inclusive and empowering, it shares similar communication shortfalls as free choice and social constructivism. In the pursuit of equality, the communication of the science behind homosexuality was oversimplified, and then extended to include other queer experiences, but only when those experiences strengthened the biodeterminist narrative. As a result, anybody within the queer community who experienced a journey different to the 'born this way' narrative was once again marginalized and silenced, this time deliberately and often by the queer community itself (Lemke, 2016). For example, those who lived a life that appeared to fit heteronormative ideals, only to come out as queer later in life, were viewed by some queer activists as inadequately fitting the narrative of 'enduring and immutable' from early in life and complicated an otherwise powerful argument for equality. High profile examples of backlashes against individuals who challenge the 'born this way' narrative can be seen in the treatment of actor Anne Heche after the ending of her relationship with comedian Ellen DeGeneres. Heche was characterized as delusional and attention-seeking, someone who was never really gay (Lemke, 2016). Actor Cynthia Nixon also faced a strong backlash from the queer community after describing her experience of choosing to be gay in an interview with *New York Times Magazine*. At least one prominent gay blogger attacked this statement as it was seen as creating an example to be used by right-wing lobbyists to deny civil rights to the queer community (Fashingbauer Cooper, 2012).

These examples demonstrate how some within the queer community view discourses that suggest sexuality is fluid as homophobic regardless of origin, because these fluid experiences echo the oft-heard phrase, 'it's just a phase'. This phrase is used to dismiss experiences of queer individuals, particularly those who identify as queer early on in life, by minimizing the experiences to something that will disappear when the queer individual becomes an adult. The alternative narrative characterizes sexuality as a fluid experience, implying a potential for anyone to experience queerness at any stage in life. This is a powerful argument for queer activism – by extending the possibility of 'queerness' to everyone, it personalizes the fight for equal rights (Lemke, 2016) and is the beginning of a more diverse conversation.

Conclusions

Science communication risks continuing the marginalization and silencing of queer people when it focuses too much on selling a simplified idea. When descriptions of what it means to be queer lack input from diverse experiences, they marginalize and silence groups within the queer community. A nuanced approach that embraces the diversity of experiences within the queer community and amplifies the vast variety of stories would ensure that no voices or experiences of queer folk are erased or misrepresented. Such an approach will also serve to support those who are struggling to understand their own experiences by better reflecting the queer community.

Science communicators have a responsibility to step up and acknowledge the nuances of what it is to be queer so that marginalized and silenced voices in the community can be heard. This cannot happen if science communicators do not own the limitations of their campaigns and agendas. In the cases of problematic science communication described in this chapter, the main issues stem from defending a particular line of rhetoric, rather than the actuality and complexity of queer lives. While this is most evident in Dr Money's willingness to actively destroy lives, it is also true for any rhetoric that fails to acknowledge the lived experience of minority members within the community. Even rhetoric deployed in an effort to assist the community needs to leave space to include the marginalized amongst us, and must be vigilant against its own excluding tendencies. A queerer science communication community, that deliberately amplifies the voices of queer people, will create a more diverse, inclusive, nuanced, and queer discourse in the public sphere.

References

Aggrawal, A. (2008) *Forensic and Medico-legal Aspects of Sexual Crimes and Unusual Sexual Practices*, Florida: CRC Press.

APA. (1952) *Diagnostic and Statistical Manual of Mental Disorders* (1st edn), Washington: APA.

APA. (1968) *Diagnostic and Statistical Manual of Mental Disorders* (2nd edn), Washington: APA.

Anon. (1974) 'Conflict Marks Conference', *The Gay Liberator*, 38(June): 2, Available from: https://jstor.org/stable/community.28037208 [Accessed 30 March 2022].

Baughey-Gill, S. (2011) 'When gay was not okay with the APA: a historical overview of homosexuality and its status as mental disorder', *Occam's Razor*, 1(2): 4–18.

Berger, V. (2005) 'List of paraphilias', *Psychologist Anytime Anywhere* [website], Available from: www.psychologistanywhereanytime.com/sexual_problems_pyschologist/psychologist_paraphilias_list.htm [Accessed 20 March 2022].

Bernstein, M. (2002) 'Identities and politics', *Social Science History*, 26(3): 531–581.

Canfield, K.N., Menezes, S., Matsuda, S.B., Moore, A., Mosley Austin, A.N., Dewsbury, B.M., et al. (2020) 'Science communication demands a critical approach that centers inclusion, equity, and intersectionality', *Frontiers in Communication*, 5: 2.

Carpenter, M. (2019) 'UN Treaty Body statements on intersex human rights in Australia', *Intersex Australia* [website] 3 October, Available from: https://ihra.org.au/35665/australia-crc-crpd-harmful-practices/ [Accessed 20 March 2022].

Chamberlain, J. (1982) '10th Annual Conference on Human Rights', *Big Mama Rag*, 10(7): 4.

Chmielewski, M., Clark, L.A., Bagby, R.M., and Watson, D. (2015) 'Method matters: understanding diagnostic reliability in DSM-IV and DSM-5', *Journal of Abnormal Psychology*, 124(3): 764–769.

Christianson, M.L. (1974a) 'Queers not allowed', *The Gay Liberator*, 40(September/October): 7, Available from: https://www.jstor.org/stable/community.28037210 [Accessed 30 March 2022].

Christianson, M.L. (1974b) 'Science wonder stories', *The Gay Liberator*, 38(June): 8, Available from: https://jstor.org/stable/community.28037208 [Accessed 30 March 2022].

Compton, S. (1974a) 'Drug experiments,' *The Gay Liberator*, 36(April): 4, Available from: https://jstor.org/stable/community.28037206 [Accessed 30 March 2022].

Compton, S. (1974b) 'Protest closes council', *The Gay Liberator*, 36(April): 2, Available from: https://jstor.org/stable/community.28037206 [Accessed 30 March 2022].

Crain, C. (2020) 'Frank Kameny's orderly, square gay-rights activism', *The New Yorker* [online] 29 June, Available from: https://www.newyorker.com/magazine/2020/06/29/frank-kamenys-orderly-square-gay-rights-activism [Accessed 20 March 2020].

Drescher, J. (2010) 'Queer diagnoses: parallels and contrasts in the history of homosexuality, gender variance, and the Diagnostic and Statistical Manual', *Archives of Sexual Behavior*, 39(2): 427–460.

Drescher, J. and Merlino, J.P. (eds) (2007) *American Psychiatry and Homosexuality: An Oral History*, New York: Harrington Park Press.

Ehrhardt, A.A. (2007) 'John Money, PhD', *Journal of Sex Research*, 44(3): 223–224.

Fashingbauer Cooper, G. (2012) 'Cynthia Nixon: I'm gay by choice', *NBC News* [online] 25 January, Available from: https://www.nbcnews.com/pop-culture/pop-culture-news/cynthia-nixon-im-gay-choice-flna1c9386406 [Accessed 24 February 2022].

Fish, J., King, A., and Almack, K. (2018) 'Queerying activism through the lens of the sociology of everyday life', *Sociological Review*, 66(6): 1194–1208.

Fitzsimons, T. (2018) '"Transsexualism" removed from World Health Organization's disease manual', *NBC News* [online] 21 June, Available from: https://www.nbcnews.com/feature/nbc-out/transsexualism-removed-world-health-organization-s-disease-manual-n885141 [Accessed 9 March 2022].

Gaetano, P. (2017) 'David Reimer and John Money gender reassignment controversy: The John/Joan case', *The Embryo Project Encyclopedia* [website] 15 November, Available from: http://embryo.asu.edu/handle/10776/13009 [Accessed 20 March 2022].

Gerson, M.N. (2015) 'BDSM versus the DSM', *The Atlantic* [online] 14 January, Available from: https://www.theatlantic.com/health/archive/2015/01/bdsm-versus-the-dsm/384138/ [Accessed 20 March 2022].

Giami, A. (2015) 'Between DSM and ICD: paraphilias and the transformation of sexual norms', *Archives of Sexual Behavior*, 44: 1127–1138.

Hollis, W. (1975) 'Sex is good', *The Gay Liberator*, 47(Winter 1975–76): 10, Available from: https://jstor.org/stable/community.28037217 [Accessed 30 March 2022].

Interact Advocates for Intersex Youth. (n.d.) 'Who was David Reimer (also, sadly, known as John/Joan)?', *Intersex Society of North America* [website], Available from: http://www.isna.org/faq/reamer [Accessed 20 March 2022].

Jennings, K. (2011) 'Fighting for Freedom in Philadelphia: Barbara Gittings, 1932–2007', *HuffPost* [blog] 25 May, Available from: https://www.huffpost.com/entry/fighting-for-freedom-in-p_b_47208 [Accessed 20 March 2022].

Johnson, D. and Rand Herron, C. (1973) 'The APA ruling on homosexuality', *The New York Times* [online archive] 23 December p 109, Available from: https://www.nytimes.com/1973/12/23/archives/the-issue-is-subtle-the-debate-still-on-the-apa-ruling-on.html [Accessed 20 March 2022].

Koh, J. (2012) 'The history of the concept of gender identity disorder (English Translation)', *Seishin Shinkeigaku Zasshi*, 114(6): 673–680.

Lemke, C. (2016) 'Not born this way: exploring an alternative discourse of surprise in once-heterosexual-identified women's narratives', *The Journal of American Culture*, 39(1): 2032.

Marmor, J., Bieber, I., and Gold, R. (1999) 'A symposium: should homosexuality be in the APA nomenclature?', in L. Gross and J. Woods (eds) *The Columbia Reader on Lesbians and Gay Men in Media, Society, and Politics*, New York: Columbia University Press.

McEwan, B. (2016) 'Emotional expression and the construction of heterosexuality', *Journal of the History of Sexuality*, 25(1): 114–136.

Meyer-Bahlburg, H.F.L. (2010) 'From mental disorder to iatrogenic hypogonadism: dilemmas in conceptualizing gender identity variants as psychiatric conditions', *Archives of Sexual Behavior*, 39(2): 461–476.

Russo, F. (2017) 'Where transgender is no longer a diagnosis', *Scientific American* [online] 6 January, Available from: https://www.scientificamerican.com/article/where-transgender-is-no-longer-a-diagnosis/ [Accessed 9 March 2022].

Sklar, J. (2018) 'The prohibitions against homosexual sex in Leviticus 18:22 and 20:13: are they relevant today?', *Bulletin for Biblical Research*, 28(2): 165–198.

Spitzer, R.L. and Fleiss, J.L. (1974) 'A re-analysis of the reliability of psychiatric diagnosis', *The British Journal of Psychiatry*, 125(1): 341–347.

State Library of NSW. (2022) 'Coming Out in the 70s', *Voices of Resistance* [online exhibition], Available from: https://www.sl.nsw.gov.au/learning/voices-resistance [Accessed 20 March 2022].

The Legacy Project. (2021a) 'Barbara Gittings – Inductee 1932–2007', *The Legacy Project* [website], Available from: https://legacyprojectchicago.org/person/barbara-gittings [Accessed 20 March 2022].

The Legacy Project. (2021b) 'Frank Kameny – Inductee 1925–2011', *The Legacy Project* [website], Available from: https://legacyprojectchicago.org/person/frank-kameny [Accessed 20 March 2022].

UN Human Rights. (2020) 'Report on conversion therapy', *Office of the United Nations High Commissioner for Human Rights* [website] May, Available from: https://www.ohchr.org/EN/Issues/SexualOrientationGender/Pages/ReportOnConversiontherapy.aspx [Accessed 9 March 2022].

Uyeda, R.L. (2021) 'How LGBTQ+ activists got "homosexuality" out of the DSM', *JSTOR Daily* [online] 26 May, Available from: https://daily.jstor.org/how-lgbtq-activists-got-homosexuality-out-of-the-dsm/ [Accessed 20 March 2022].

Walker, J. (2004) 'The death of David Reimer: a tale of sex, science and abuse', *Reason* [online] 24 May, Available from: https://reason.com/2004/05/24/the-death-of-david-reimer/ [Accessed 20 March 2022].

Weber, S. (2021) 'What's wrong with be(com)ing queer? Biological determinism as discursive queer hegemony', *Sexualities*, 15(5/6): 679–701.

Whalen, K. (n.d.) '(In)Validating transgender identities: Progress and trouble in the DSM5', *The National LGBTQ+ Taskforce Online* [website], Available from: https://www.thetaskforce.org/invalidating-transgender-identities-progress-and-trouble-in-the-dsm-5/ [Accessed 9 March 2022].

World Health Organization. (1965) *International Classification of Diseases (ICD-8a)*, World Health Organization.

Zucker, K.J. and Spitzer, R.L. (2005) 'Was the gender identity disorder of childhood diagnosis introduced into DSM-III as a backdoor maneuver to replace homosexuality? A historical note', *Journal of Sex & Marital Therapy*, 31(1): 31–42.

3

Queer Interests in Technology and Innovation Discourse

Tara Roberson

Technology plays a significant role in our lives. Dominant narratives for technology might centre on the benefits, but for queer people the opportunities afforded by technology are counterbalanced by the consequences.

As a younger queer person, access to the information and social networks available through the Internet played a key role in my self-discovery. I vividly remember searching the web using different queer identities as keywords (lesbian, gay, bisexual, queer) while completing online web surveys which promised to concretely label my identity and hesitating over the selection of 'who I was interested in' while adding to my Facebook profile. Through the family computer and ADSL, I possessed an avenue for finding out what it meant to be queer while existing in a largely heteronormative world. This combination of technology and networks meant access to social groups, ways to find other people like me. Today it still serves that purpose by helping me communicate with and join people who are organizing social movements for LGBTIQA+ communities. Yet, my public 'out' presence also poses risks that I am mindful of, particularly when it comes to global travel (Equaldex, 2022; ILGA, 2020).

The pervasive and powerful presence of different technologies can be both a boon and a problem. This means science communicators who work with and through technologies need to be critically awake to the implications of their use. Here, I explore the impact technologies have on queer people and voices. I consider how queer voices might be added into the technology and innovation creation and use discourse. The underlying motivation of this work parallels Emily Dawson's (2019) eloquent articulation as to why science communication should learn to value differences instead of minimizing and erasing them. In her book on 'race' and informal science learning, Dawson says that as people who feel included in science and technology worlds, science communicators and associated professionals 'have a responsibility to centre social justice in our work. If we do not, we have to ask serious questions about whether we are happy to reproduce advantages for dominant groups at the expense of the minoritised' (Dawson, 2019, p 159). Here, I apply this lesson to technology and innovation discourse and consider what communication looks like when equity, diversity, and inclusion are centred, rather than given lip service (Yep, 2003).

Defining the conditions of queer existence

Knowledge, products, and policies that do not reflect on LGBTIQA+ people as adopters or creators can be both actively and passively harmful. The European Digital Rights (EDRi) organization argues that digital technologies can be both empowering and problematic with impacts in either direction chiefly influenced by whether communities are involved during design and creation stages (EDRi, 2019). Researchers from fields that include science and technology studies, human-computer interaction, and technical communication have conducted research along these lines that science communication could benefit from.

For instance, scholars from science and technology studies have traced how gender and gendered norms influence and define scientific insights (Rosner, 2019). Studies in this discipline have considered how gender norms influence participation in technology design and change narratives to enforce absence of other identities. For example, in IT environments these norms can lessen the visibility of women's work in software development (Couture, 2019). Meanwhile, researchers in human-computer interaction have considered the challenges faced by LGBTIQA+ people in using technologies. These include technologies that reject the existence of gender diverse people, such as binary gender options of 'male' and 'female' in webforms. Studies in this field have opened up conversations around the need for co-design with these communities. In Brazil, one of these projects focused on co-designing social technology (a mobile app) to support LGBT+ people working against prejudice in sometimes hostile environments. Finally, in technical communication, some researchers have presented a theoretical framework for queer, tactical technical communication using the example of working outside of – and resisting – the cis-heteronormative project (Edenfield et al, 2019). Their framework uses the example of individual-created guides to 'do-it-yourself' (DIY) hormone replacement therapy for gender diverse people stymied by medical services.

A key theme to these threads of research is how the design of technology defines the possibility and boundaries of queer existence. These boundaries can be physical and virtual. Human-computer interaction researchers have examined how virtual worlds created within computer games can provide safe havens for oppressed people (Cabiria, 1970; Döring, 2009; Collister, 2014). Virtual worlds can be important places for LGBTIQA+ participants and yet they exist under the threat of closure (Ducheneaut and Moore, 2007). Take for example massive multiplayer roleplaying games such as *World of Warcraft*, which can explicitly and implicitly allow or ban the presence of queer characters and players. *World of Warcraft* provides a vast online environment for players to pursue individual goals and meet with other participants. To maintain queer spaces within *World of Warcraft*, LGBTIQA+ players have engaged in virtual protests and other forms of resistance to make the game friendly to all genders. Players have also created venues for online connection between queer individuals outside of the traditional bounds of the game (McKenna and Chughtai, 2020). These players are pushing back on the conditions that define queer existence within these domains and, in some cases, creating a more inclusive, diverse online world.

Furthering this focus on protecting and supporting queer existence, researchers have sought to understand how technology can be designed to support the needs of trans, non-binary, and other gender diverse people. One attempt from Haimson and colleagues (2020) considers how technological solutions could help address challenges faced by trans people and communities. Building on participant design sessions with trans participants, they call for the development of 'trans technology'. Trans technology can include hardware (virtual reality glasses) or software (smart phone applications) that contain features (such as 'changeability, network separation, and identity realness') to support users' gender transitions and affirmations (Haimson et al, 2020). These technologies can celebrate and showcase trans identities and experiences. They can also improve safety, provide resources, and build community for users. While those authors have kept the definition of the phrase purposefully broad, a concrete example might include smartphone apps that enable secure and private connections between trans and gender diverse people.

Anti-trans technology, on the other hand, is technology that reinforces the gender binary and, therefore, poses risk of harm and oppression to the people who are misclassified or who fear misclassification (Edenfield et al, 2019). This technology defines queer existence narrowly by excluding gender diverse peoples. One example of anti-trans technology is a subfield of facial recognition: automatic gender recognition systems (Keyes, 2018). Automatic gender recognition systems identify gender using information from sources that include photographs and videos. They commonly exclude and pose risks to trans and non-binary peoples due to the use of binary classification systems and cis-heteronormative assumptions around the relationship between gender, physical characteristics, and gender expression.

Other uses of facial recognition software pose the risk of harm to a variety of LGBTIQA+ people. For example, researchers at Stanford University claimed to have created software that could identify sexual orientation. The research team claimed to have created an algorithm that could distinguish between heterosexual and homosexual men and women after teaching the system to recognize people using a database of photos from a dating site of over 14,000 white Americans. As legal researcher Anat Lior notes (2018), this technology poses at least three legal issues with the use of artificial intelligence (AI) as a 'technological gaydar'. Those are: first, privacy infringement of people via the use of photos which were not specifically provided to the research team; second, the potential use of the results as evidence in countries which criminalize homosexuality; and third, the loss of individual autonomy through the spread of biometric tools.

One aspect of digital technology that already pervades our lives – social media – has been labelled 'effectively unsafe' for queer users by LGBTQ media advocacy group GLAAD (2021). Specific concerns listed in GLAAD's 2021 *Social Media Safety Index* report included hate speech and harassment with inadequate content moderation, problematic algorithms, and discriminatory AI. Social media companies have taken public steps (including statements and the establishment of research teams) to mitigate the impact of their platforms on multiple marginalized communities. However, more action – taken in partnership with these communities – is needed if social media is to change for the better so that it can support and protect LGBTIQA+ peoples in place of harming them.

In these examples, we have so far glimpsed instances from the areas of online gaming, automated classification systems, and social media in which LGBTIQA+ lives and experiences are shaped and defined by the constraints built into platforms. Users of these platforms have, in some cases, asserted their own power by resisting the cis-heteronormative restrictions placed on them. In other instances, those normative influences pose risks to queer communities by potentially creating and amplifying opportunities for discrimination and harassment or, indeed, excluding their existence entirely. Now, I turn to stories that provide us with an alternative approach, one that centres the role of LGBTIQA+ actors in the research and development process and considers the role of technology and innovation in actively supporting LGBTIQA+ people.

Technology and innovation as emancipation

In 2015, an industry-based panel discussion on technology and development highlighted the complex combination of opportunities and challenges afforded by technology and the tech industry for LGBTIQA+ people (Raftree, 2015). Queer-identified industry professionals highlighted the heightened privacy needs of LGBTIQA+ communities, noting that data collected about individuals needed to be anonymous and/or kept safe from malicious parties. This was particularly important in countries where governments might monitor or otherwise trace online activities. Working within the technology industry as LGBTIQA+ people in international organizations with offices located in different countries (particularly countries in which their queer identity might be outlawed or otherwise pose a risk to their safety) was another challenge. Speakers on the panel outlined the steps they had taken to conceal their identity because of the location of their work and the ongoing seepage of information between online professional and personal profiles, which heightened potential risks and sometimes limited career choices.

An organization based in the United States has adopted these privacy-focused concerns as a key pillar in its ongoing research and advocacy work. The LGBT Technology Partnership and Institute (LGBT Tech) aims to support LGBTQ+ individuals while educating organizations and policymakers around their unique needs when it comes to technology (Gutierrez, 2021). Included among their focus areas are the need for encryption and privacy as well as revision of filtering technologies that block LGBTIQA+ content on public networks and improvement of telehealth legislation (LGBT Tech, 2022). They also highlight the 'digital divide' in relation to the significance of connectivity, especially for rural and homeless queer communities (Daniels and Gray, 2014), and risks posed by emerging technologies, such as unregulated use of facial recognition technology (Gutierrez, 2019).

In a 2012 report on homelessness, LGBT Tech found that the LGBTQ+ population was disproportionately affected by homelessness and that homeless LGBTQ+ people faced specific, heightened risks on the streets (Hackl, 2014). As part of an effort to support this community, LGBT Tech established *PowerOn*, an initiative that supports at-risk and under-served queer individuals by providing technology (including phones and software licenses) at no cost through partner centres in the United States and Puerto Rico (PowerON, 2021). Access to technology is designed to help individuals keep in touch with support networks and critical services, gain access to healthcare information and services, and improve personal safety.

These developments recognize the potential role of technology in the lives of LGBTIQA+ people, particularly in cases where it provides mechanisms for accessing information, joining vital social networks, and coordinating social movements. However, these same platforms for engaging with one another present risks and obstacles where security is uncertain and access to resources is uneven. The rise of LGBTIQA+ technology advocacy and industry groups is one step in redressing this imbalance and underscores the need for queer voices in this space.

Resistance as participation in technology and innovation discourse

The increasing emphasis on robust participation and inclusion emerging in science communication has also occurred in other fields, including technical communication and user experience design (Ahmed et al, 2020; Devito et al, 2021). These researchers have focused on understanding and engaging with resistance – for instance, tactical technical communication practices in which individuals 'usurp' technology to achieve their own goals (Kimball, 2006). One example (previously mentioned) is the creation and sharing of guides on do-it-yourself hormone replacement therapy. These guides were written by and for gender diverse people who have been stymied by medical, legal, and insurance institutions (Edenfield et al, 2019). Such examples of resistance can provide valuable indications of where design fails to align with the interests of users.

For a technology-focused example, consider the design of social media which incorporates a default setting of publicness based on the assumption that a person's offline and online lives work simultaneously or in parallel (Cho, 2018). This public setting might suit individuals whose selves and bodies navigate the offline world easily. However, the setting works against those who do not fit dominant social norms and institutions (Green, 2020). By extension, public health campaigns and organizations that then adopt social media to disseminate information run the risk of reinforcing the existing social orders and inequities the public setting assumes while failing to reach – and potentially alienating – the communities they intend to assist.

In an example of this, Green (2020) interviewed queer individuals living with HIV to gain feedback around an initiative they designed and implemented with a public health organization in the United States. The public health initiative focused on HIV health education via digital technology. Green describes the design choices the team made, which were based around a set of assumptions. These assumptions were: that young people would receive health information on HIV via social media and digital technologies; social media would provide a neutral and useful platform for sharing information; and a deficit model influenced belief that 'redressing health disparities required delivering information to people living with HIV' (Green, 2020, p 2). Contrary to these initial assumptions, Green found that participants were distrustful of social media – chiefly because they were highly aware of the need to control what information was known by whom and where details might be shared, or become insecure, online. Participants also strongly indicated that rather than seeking technology that would passively deliver information, they desired technology that would empower them and build a community with a space for sharing stories and learning from one another. Critiques of the programme offered through these interviews explained early user resistance, produced a 'lens to understand how digital systems privilege some bodies and experiences while dismissing others', and a framework for understanding how offline inequities shape design and use of technology (Green, 2020, p 12).

In the landscape of social justice in technical communication, resistance has become a key component of the focus on 'ethical action within, rather than efficient consumption of, emerging technologies' (Opel and Rhodes, 2018, p 74). Resistance challenges assumptions about – in this case – sexuality, desire, companionship, and bodies (Edenfield et al, 2019). It is also a way of shifting the focus of communication or use of technology from industry-led goals to the promotion of the 'values of individuals whose lives intersect with technologies' (Green, 2020, p 3). Queer-Alt-Delete is one example of this approach, created by feminist researcher Andie Shabbar (2018). Queer-Alt-Delete substitutes 'queer' for 'control' in the function Control-Alt-Delete, which users enact to interrupt functions on a computer. The insertion of queer into this function is Shabbar's response to biometric recognition technologies which 'aim to fix the body within rigid identity categories' in environments such as airports. These technologies tend to mark trans bodies as 'biometric failure' and enforce additional scrutiny of those who do not confirm to cisgender norms. In Queer-Alt-Delete, Shabbar introduced deliberate glitches (or errors) in code to damage digital image files. The product of these errors is a warped image. Shabbar's intent in this process is to resist binary labels within technologies like facial recognition technology. By interrogating and interrupting the glitches that trans bodies introduce in biometric surveillance technology, their project aims to expose the vulnerabilities and fallibility of recognition technologies.

Instances of resistance provide insight into how technology is experienced by people who are part of underrepresented groups. In the two examples considered here, technologies were designed and used based on assumptions around how queer people navigate digital technologies and the relationship between our bodies and our genders. In the case of the public health initiative, these assumptions hindered the success of the programme because it did not connect with the needs of queer people with HIV. Meanwhile, in the second example, the use of biometric surveillance technologies enforces white and cisgender norms upon people being monitored. Shabbar's project, then, works to disrupt those norms and illuminate the 'racist, ablest, and transphobic schemas' built into biometric technologies from the very beginning of design processes.

Lessons for queering science communication

Emphasizing equity and inclusion with science and innovation does not only benefit people from underrepresented communities. Research into science and innovation tells us that diversity supports innovation (Hofstra et al, 2020). Historically underrepresented groups produce higher rates of scientific novelty, though their contributions can be devalued and discounted. Sociology researcher Bas Hofstra and colleagues call this the 'diversity-innovation paradox'. People from underrepresented groups bring diverse perspectives to their scholarly work and these perspectives allow them to draw novel relations between concepts and ideas, yet despite this capacity for innovation, the careers of these researchers do not always benefit. On the flipside of this discussion, when technology design does not deliberately incorporate diverse voices during early design and development stages, the technology created will likely meet the needs of the few rather than the many (Parthasarathy, 2020).

Diversity within science and innovation processes is vital. Communication plays a central role here – not just in the sense of information transfer, but also in terms of how we make sense of new developments and fit them into the world. When we consider this in conjunction with the recent push for inclusive science communication, a focus on communication helps us examine whose voices are and are not heard.

Science communication works as a boundary space for interaction where different stakeholders can meet and interact with empathy (Horst, 2021). These interactions are difficult to make real when interactions easily default to deficit-style approaches (Dawson, 2019; Metcalfe, 2019). The challenge, then, is to determine how equity, diversity, and inclusion can be embedded in communication processes. In other words, how can we engage with care and focus on previously neglected people and ideas (de la Bellacasa, 2011)? If we understand and engage with these ideas early on, then we are more prepared for what it means to be inclusive-by-design when it comes time to put them into practice.

We know there is a gap in science communication when it comes to engaging with queer people and queer experiences (Roberson and Orthia, 2021). Consider the role science communication played in the COVID-19 pandemic, helping people understand the coronavirus and communicate about behaviours that could minimize its spread. Research in the field since 2019 has examined a range of topics, including misinformation, the impact of fake news, and how we talk about vaccination. One gap for review amidst this activity is how public health strategies impact marginalized communities, including LGBTIQA+ people. Specific examples include the use of gender-based segregation during public lockdowns in Panama and the impact on the gender diverse community; use of geolocation in tracing apps in countries such as South Korea (which, in this case, fuelled fears of being publicly outed); and the 'unique vulnerabilities' of LGBTIQA+ people in relation to both contracting COVID and experiencing negative impacts from social distancing measures, including lockdowns (UN Human Rights, 2021). Recognition of these impacts on LGBTIQA+ people prompted the United Nations to produce guidelines in late 2020 that focused on how to shape public health initiatives by, for instance, working with LGBTIQA+ communities in early design stages. The pandemic has had significant, ongoing, and sometimes disproportionate impacts for LGBTIQA+ communities. It serves as a timely reminder of the importance of applying a queer lens if we seek to centre social justice in science communication initiatives and research.

In this chapter, we have glimpsed how technologies can restrict, govern, and enable queer lives. Building on this, how then should science communicators queer technology and innovation discourses? A case study on trans technology that was briefly discussed earlier may provide some insight.

In 2019, human computer interaction researcher Oliver Haimson and colleagues (2020) organized a series of participatory design workshops with trans and non-binary people to discuss how technology might be designed to meet community needs. These workshops generatively created a list of challenges faced by the community. This list included challenges of accessing healthcare and other resources, facing violence, building community, and ensuring personal safety. Using this list, the workshop participants then created ideas for how technology might address or alleviate these challenges. The participatory process produced technology concepts within four themes: technology for changing bodies, technologies for changing appearances and gender expressions, technologies for safety, and technologies for finding resources. In a blog post reflecting on the experience, Haimson writes that many of the designs 'included community-based aspects in serious ways' (Haimson, 2020). In other words, participants of the design workshops prioritized connection between people over and beyond designing tools for singular, individual use. Haimson concludes that this means trans technology must be designed for the trans community as well as individuals to be successful.

This project is a practical example from which the science communication field might learn as we strive to improve the inclusiveness of our own practices. In particular, the success of the project was based on how well the researchers listened to and learned from underrepresented communities.

Queering technology and innovation discourses requires multiple layers of action and reflection from science communicators and other actors. We need to reflect on how the assumptions we make about people (their sexuality, gender, bodies, and so on) influence how we design parameters for our projects. These assumptions also influence how we anticipate engagement with and through technology. Keeping our eyes open to acts of resistance may help disrupt preconceptions. Providing formal and informal feedback loops in design processes through community engagement, interviews, and other mechanisms would also assist here.

References

Ahmed, A.A., Kok, B., Howard, C., and Still, K. (2020) 'Online community-based design of free and open source software for transgender voice training', *Proceedings of the ACM on Human-Computer Interaction*, 4(CSCW3): 258.

Cabiria, J. (1970) 'Virtual world and real world permeability: transference of positive benefits for marginalized gay and lesbian populations', *Journal of Virtual Worlds Research*, 1(1): 1–13.

Cho, A. (2018) 'Default publicness: queer youth of color, social media, and being outed by the machine', *New Media & Society*, 20(9): 3183–3200.

Collister, L.B. (2014) 'Surveillance and community: language policing and empowerment in a World of Warcraft guild', *Surveillance and Society*, 12(3): 337–348.

Couture, S. (2019) 'The ambiguous boundaries of computer source code and some of its political consequences', in J. Vertesi and D. Ribes (eds) *digitalSTS: A Field Guide for Science & Technology Studies*, Princeton: Princeton University Press, pp 136–156.

Daniels, J. and Gray, M.L. (2014) 'Vision for inclusion: an LGBT broadband future addresses unique technology needs of the community', *LGBT Tech* [online] 26 October, Available from: https://www.lgbttech.org/_files/ugd/699ad7_b0219ea9c8804c05a03d95ca7f911f78.pdf [Accessed 20 March 2022].

Dawson, E. (2019) *Equity, Exclusion and Everyday Science Learning: The Experiences of Minoritised Groups*, Oxon: Routledge.

de la Bellacasa, M.P. (2011) 'Matters of care in technoscience: assembling neglected things', *Social Studies of Science*, 41(1): 85–106.

Devito, M.A., Walker, A.M., and Fernandez, J.R. (2021) 'Values (Mis)alignment: exploring tensions between platform and LGBTQ+ community design values', *Proceedings of the ACM on Human-Computer Interaction*, 5(CSCW1): 88.

Döring, N.M. (2009) 'The internet's impact on sexuality: a critical review of 15 years of research', *Computers in Human Behavior*, 25(5): 1089–1101.

Ducheneaut, N. and Moore, R. (2007) 'The social side of gaming: a study of interaction patterns in a massively multiplayer online game', *Proceedings of the 2004 ACM Conference on Computer Supported Cooperative Work (CSCW'04)*, New York, pp 360–369.

Edenfield, A.C., Holmes, S., and Colton, J.S. (2019) 'Queering tactical technical communication: DIY HRT', *Technical Communication Quarterly*, 28(3): 177–191.

EDRi. (2019) 'The digital rights of LGBTQ+ people: when technology reinforces societal oppressions', *EDRi* [website] 17 July, Available from: https://edri.org/our-work/the-digital-rights-lgbtq-technology-reinforces-societal-oppressions/ [Accessed 20 March 2022].

Equaldex. (2022) 'LGBT Equality Index', *Equaldex* [website], Available from: https://www.equaldex.com/equality-index [Accessed 20 March 2022].

GLAAD. (2021) *Social Media Safety Index*, USA: GLAAD.

Green, M. (2020) 'Resistance as participation: queer theory's applications for HIV health technology design', *Technical Communication Quarterly*, 30(4): 331–344.

Gutierrez, C. (2019) 'Unregulated facial recognition technology presents unique risks for the LGBTQ+ community', *TechCrunch* [online] 29 June, Available from: https://techcrunch.com/2019/06/29/unregulated-facial-recognition-technology-presents-unique-risks-for-the-lgbtq-community/ [Accessed 20 March 2022].

Gutierrez, C. (2021) *Declaration of LGBT Technology Institute in Support of Plaintiff's Motion for Preliminary Injunction*, Submission to United States District Court for Western District of Texas Austin Division.

Hackl, A.M. (2014) 'Helping homeless youth stay connected: LGBT Tech Connect 4 Life Program and Research', *LGBT Tech* [online] 6 Sept, Available from: https://www.lgbttech.org/_files/ugd/699ad7_40e710811f30494f96f0282b50c6a7ad.pdf [Accessed 25 March 2022].

Haimson, O.L. (2020) 'Defining and designing trans technologies', *Medium* [online] 29 January, Available from: https://medium.com/@haimson/defining-and-designing-trans-technologies-90cc77c8cc0b [Accessed 20 March 2022].

Haimson, O.L., Gorrell, D., Starks, D.L., and Weinger, Z. (2020) 'Designing trans technology: defining challenges and envisioning community-centered solutions', *Proceedings of the 2020 CHI Conference on Human Factors in Computing Systems*, Honolulu, USA.

Hofstra, B., Kulkarni, V.V., Munoz-Najar Galvez, S., He, B., Jurafsky, D., and McFarland, D.A. (2020) 'The diversity-innovation paradox in science', *Proceedings of the National Academy of Sciences*, 117(17): 9284–9291.

Horst, M. (2021) 'Science communication as a boundary space: an interactive installation about the social responsibility of science', *Science, Technology, & Human Values*, published online 5 April 2021, doi:10.1177/01622439211003662.

ILGA. (2020) 'Maps – sexual orientation laws,' *ILGA World* [website] December, Available from: https://ilga.org/maps-sexual-orientation-laws [Accessed 20 March 2022].

Keyes, O. (2018) 'The misgendering machines: trans/HCI implications of automatic gender recognition', *Proceedings of the ACM on Human-Computer Interaction* 2(CSCW): 88.

Kimball, M.A. (2006) 'Cars, culture, and tactical technical communication', *Technical Communication Quarterly*, 15(1): 67–86.

LGBT Tech. (2022), 'Establishing a Pathway for Comprehensive Telehealth Reform', *Correspondence to United States of America Congress* [online], 31 January, Available from: https://www.lgbttech.org/_files/ugd/699ad7_40f0371df1f34034979c642f71e5af7a.pdf [Accessed 25 March 2022].

Lior, A. (2018) 'The technological "Gaydar" – The problems with facial recognition AI', *Yale Journal of Law & Technology* [blog] 11 January, Available from: https://yjolt.org/blog/technological-gaydar-problems-facial-recognition-ai [Accessed 20 March 2022].

McKenna, B. and Chughtai, H. (2020) 'Resistance and sexuality in virtual worlds: an LGBT perspective', *Computers to Human Behaviour*, 105: 106199.

Metcalfe, J. (2019) 'Comparing science communication theory with practice: an assessment and critique using Australian data', *Public Understanding of Science*, 28(4): 382–400.

Opel, D.S. and Rhodes, J. (2018) 'Beyond student as user: rhetoric, multimodality, and user-centered design', *Computers and Composition*, 49: 71–81.

Parthasarathy, S. (2020) 'More testing alone will not get us out of this pandemic', *Nature*, 585(3 September): 8.

PowerON. (2021) *Impact Report FY2020*, LGBT Tech, Available from: https://www.poweronlgbt.org/_files/ugd/1b643a_2fbd50061edd4225a7cb79fb19f89f38.pdf [Accessed 20 March 2022].

Raftree, L. (2015) 'An understanding of LGBTQI rights and technology for development', *Technology Salon* [online] 4 May, Available from: https://technologysalon.org/an-understanding-of-lgbtqi-rights-and-technology-for-development/ [Accessed 20 March 2022].

Roberson, T. and Orthia, L.A. (2021) 'Queer world-making: a need for integrated intersectionality in science communication', *Journal of Science Communication*, 20(1): C05.

Rosner, D.K. (2019) 'Introduction: gender', in J. Vertesi and D. Ribes (eds) *digitalSTS: A Field Guide for Science & Technology Studies*, Princeton: Princeton University Press, pp 77–80.

Shabbar, A. (2018) 'Queer-Alt-Delete', *Women's Studies Quarterly*, 46(3 & 4): 195–212.

UN Human Rights. (2021) 'An LGBT-inclusive response to COVID-19', *Office of the High Commissioner for Human Rights* [website], Available from: https://www.ohchr.org/EN/Issues/SexualOrientationGender/Pages/COVID19LGBTInclusiveResponse.aspx [Accessed 4 February 2022].

Yep, G.A. (2003) 'The violence of heteronormativity in communication studies', *Journal of Homosexuality*, 45(2–4): 11–59.

Practice Spotlight: All We Need Is ... The Endosymbiotic Love Calendar

*Annalaura Alifuoco, Natalie E.R. Beveridge,
Yasmine Kumordzi, and Hwa Young Jung*

Endosymbiotic Love Calendar 2021 is an art-meets-science project supported by Arts Council England and the Microbiological Society and was completed in 2020 (Endosymbiotic Love Calendar, 2022). The creative multidisciplinary team included researchers and practitioners: artists, filmmakers, microbiologists, a designer, a physician, a neuroscientist, and an immunologist. We considered 12 different microorganisms, studying how their complex and dynamic forms revealed them as human companion species in the world. The objective was to explore identity, performativity, and social justice through a queer ecology that reimagined aesthetically these symbiotic relationships.

This initiative aimed to render these transdisciplinary and trans-species alliances accessible through an everyday household object. We chose the domesticity of a wall-mounted calendar as a familiar space to reflect on unfamiliar relations. The calendar featured theatrical and curious characters derived from our collective deviant adventures in microbial cultures: one microorganism for every month of 2021. As we explored the physical, emotional, and political relationships between macro- and micro-organismic life, the images began to embody and enact a playful form of storytelling.

Within this framework, the meanings of queerness were practiced in the broadest sense. The collective comprised members identifying as lesbian, gay, bi+, trans, genderqueer, and non-binary, alongside non-rainbow identities. More than that, we regarded the qualities of microorganisms themselves as being radically queer in their 'exhilarating' and 'perverse' ways of existing and relating with the environment and other organisms. Their systems of reproduction, plasticity, and sociality allow for an exploration of more-than-human bodies and symbiotic kinships. Their transmutability and adaptability reorient the heteronormative biases in dominant evolutionary theories.

For example, in response to their environment microorganisms can change physical form, adjust their metabolism, biochemical pathways, and even their reproduction methods. They form networks of care and allegiances, sharing resources across species, which includes trees in the case of underground fungal webs. Microorganisms share genetic material within their species and across others, generating richness and depth. Their ways of being cut across binary, colonial imperialist, capitalist, and anthropocentric notions of human life. In contrast, microbes are mutualistic; we do not believe the concept of 'othering' exists in microorganisms.

In the calendar, these characteristics and qualities are activated through the discursive and aesthetic explorations that inform the monthly images and descriptions collated in the form of dating profiles. For example, the Tulip breaking virus page (Figure Sp2.1) features an image of a fuzzy portrait of textured Y-front briefs showing a section of skin with stretchmarks as the physical impression of the effect of this organism on the flower's petals. This is accompanied by the blurb:

> Infectious, striking type looking for a sapiosexual to indulge our morbid love for literature, history, and death. Once you get to know us, the memory will remain etched on your skin. Beauty is in the bulb of the beholder. Let's ignite our two-lips chasm and burn out in scarlet flames of passion.

The overarching intention was to nurture an understanding of embodied relations, overriding notions of genus, species, and identity, in compassionate struggles for love and life. We hoped to produce a playful sense of synergistic relations and associations where we can all learn, depend on, and become a part of each other – something more than the sum of our separate entities. This was the inspiration behind the title – *Endosymbiotic Love* – Endo meaning *within* and symbiotic meaning *the relationship between* two species: where positive, negative, and neutral fall into the mix of homeostasis.

In the end, this project was pulled together across three different countries during the new critical relations emerging around the COVID-19 crisis, proving just how microbial and other life are intricately enmeshed. Like all organisms we had to adapt to the new environments and demands, in the urgent need of mindful relations suggestive of what is most essential: love, kinship, and care as ways of knowing, relating, and being.

Figure Sp2.1: Tulip breaking virus, 2020

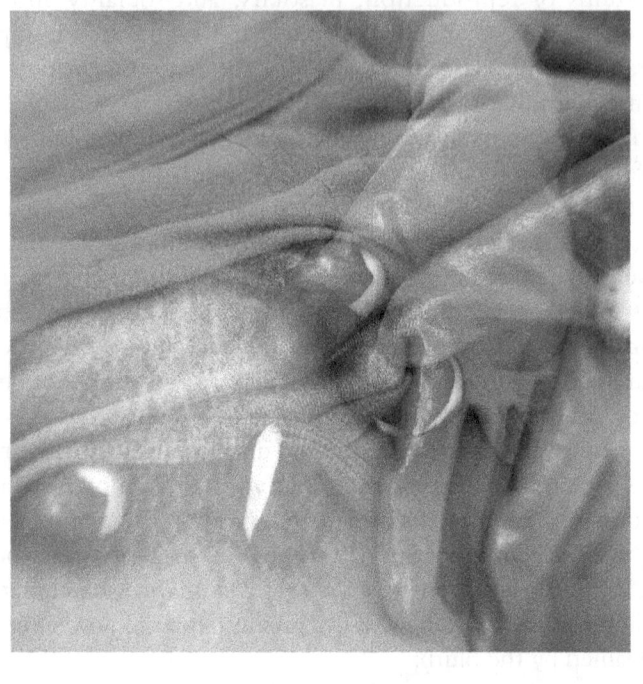

Source: Annalaura Alifuoco, Lee Haines, Fabiola Paz

References

Endosymbiotic Love Calendar. (2022) 'About,' *Endosymbiotic Love Calendar* [website] January, Available from: https://endolove.slyrabbit.net/about/ [Accessed 20 March 2022].

Practice Spotlight: GENDERS: Shaping and Breaking the Binary, an Exhibition at Science Gallery London

Helen Kaplinsky and Jessie Krish

'GENDERS: Shaping and Breaking the Binary' (2020) was an interdisciplinary exhibition and public programme at Science Gallery London (SGL) on themes of queer ecology, gender expression, roleplay, and reproduction (Science Gallery London, 2020). The programme engaged a multidisciplinary entanglement of social and biological research.

It comprised two research projects and 18 artworks – nine were new commissions and four were coproduced with researchers and community groups. Artists, activists, and researchers were selected through open call and invitation. A youth board and advisors consulted on the selection and development of projects. SGL is based within the healthcare campus of King's College London (KCL) and the production of artworks was enabled by relationships with King's researchers. The notion of 'queering hospitality' emerged through reflective conversations with contributors who commented on the misuse, or 'queering' of institutional resources, as well as conflicting agendas that question how scientific knowledge is shaped. Here we recount a few of the contributions to GENDERS.

Skin Flicks (invasive species): artist Adham Faramawy reflected upon the encounter between queer narratives in their artwork and the gallery context, adjacent to a medical school and hospital. The protagonist in *Skin Flicks (invasive species)* (2020) experiments with what is known as 'hormone hacking'. Used by bodybuilders to boost performance, accessing hormones in semi-illegal ways is also a common experience for members of the trans community seeking to achieve physiological changes as part of gender affirmation. When interviewed,[1] Faramawy said, "This story talks about certain forms of unlegislated experiments with the body that are outside the practice of the medical profession. It became interesting to install that part of it in a space ... where doctors are trained." Faramawy highlights how SGL's institutional embeddedness provided a meaningful opportunity to challenge norms and expand conceptions of medical practice.

'On being allergic to onions' ... **we read Susan Leigh Star:** Anne Pollock, Professor of Global Health and Social Medicine at KCL and a curatorial advisor commented on her experience contributing to the new commission by artist Nina Wakeford *'on being allergic to onions' ... we read Susan Leigh Star* (2020). She visited a neuroendocrinology lab together with the artist and a group of collaborators who are drag kings, reporting that "entering a lab in the company of such an eclectic group of queers, spurred a refreshingly lively set of conversations, allowing distinctive engagement with vital questions of how gender is operating and operationalized in biomedical research – it was extraordinary". The tour from scientists sought to equip the drag kings with an understanding of the vernacular of lab researchers, which went on to inform the characters they performed in the commissioned artwork. In a reciprocal gesture of hospitality, the drag kings hosted a drag workshop for staff and students at King's.

Recombinant Commons: Mary Maggic's installation *Recombinant Commons* (2019) included a model of a river in Indonesia made from agar, a medium used in laboratories to facilitate bacterial growth. The river contains chemicals from pollution that behave like the hormone estrogen, producing queer mutations. Maggic's agar river hosted bacteria in the gallery and was embedded with petri dishes containing bioremediating fungi. A form of queer life that reproduces asexually, fungi are also capable of breaking down so-called 'toxic' chemicals. A lab at King's was repurposed to cultivate the fungi and Judit Agui, production assistant, described how on transferring it to the gallery, "the fungi grew a lot and broke the seals of the petri dishes and started colonising the agar". The fungi's extraordinary growth, looked upon with excitement by the artist, exceeded the boundaries of visitor health and safety. Agui's job moved from creating a hospitable infrastructure to containing the live organism. The institutional limit to queer hospitality was drawn at the potential contamination of visitors, raising the question central to the artwork: what counts as 'toxic'?

What does gender mean to you? A research project in the GENDERS exhibition asked young people what gender means to them through drawing and text. Early analysis of the results suggests that some young people were not able to imagine life outside of a binary, whilst many others were increasingly open about gender, recognizing it as being fluid, even if at times it feels restrictive. Project co-author Adam Shepherd (2021) observed, 'the question remains as to how these new and shifting models of gender, that recognize everyone's myriad and diverse experiences, will enter into healthcare provision.' Shepherd's insight, that there is a temporal lapse, causing tension between the fluidity of emerging ideas and fixed institutional practices, provides a wider frame for the scenarios explored by way of queering hospitality.[2]

As we described earlier, the GENDERS exhibition became an occasion for human and non-human queer bodies to renegotiate the institutional norms of university, hospital, and gallery, challenging the fixed binary categories that often characterize learning and understanding in research and medical communities and for audiences. Head of Programming Jen Wong observed that, distinctive from other SGL seasons, the relationships between artists and research collaborators in GENDERS "were less transactional and instead made space for complexity". Highlighting GENDERS as a project from which SGL continues to draw learning, Wong suggests that science communication methodologies become richer and more relevant when, alongside academic multidisciplinarity, they integrate knowledge drawn from queer lived experience.

Notes

[1] Quotes from collaborators used with permission from interviews by Krish and Kaplinsky, all 2021.
[2] Research project co-authored by Benjamin Hanckel and Adam Shepherd.

References

Science Gallery London. (2020) 'GENDERS: Shaping and Breaking the Binary,' *Science Gallery London* [website] February, Available from: https://london.sciencegallery.com/genders [Accessed 20 March 2022].

Shepherd, A. (2021) 'International transgender day of visibility: why visibility in healthcare matters', *Diversity Digest: Equality, Diversity & Inclusion at King's College London* [blog] March 29, Available from: https://blogs.kcl.ac.uk/diversity/2021/03/29/international-transgender-day-of-visibility-why-visibility-in-healthcare-matters/ [Accessed 5 October 2021].

Teaching Notes for Part I

At the end of Parts I–IV of the book, we bring together the part's key themes through questions and activities for university teachers to use with students. While teaching oriented, this content could also be readily re-used for other purposes, for instance as points for reflection in the design of science communication initiatives.

If you choose to use these notes in teaching, please keep in mind that any LGBTIQA+ students in your class should not be made to feel like an object of study. This applies whether they are out about their LGBTIQA+ status or not.

It is also best if you begin by situating your own positionality as much as you feel comfortable when introducing these activities. That is, state your pronouns and your affiliation with the queer community and with queer support networks at your institution.

It may also be helpful to provide links to student support services and other avenues that LGBTIQA+ students will find useful and empowering.

Questions posed by Part I:

- **Science communication models:** a one-way model of transferring information from scientists to a broader audience is often criticized within science communication. When considering the ways that medical scientists and psychologists have talked about queer people, what are some of the problems with a one-way transmission model? What kinds of communication model would be more appropriate for ensuring LGBTIQA+ people remain empowered in such situations? In what ways have LGBTIQA+ people 'spoken back' to science?
- **Sources of expertise:** LGBTIQA+ people have been the subject of scientific studies for centuries, and the scientists who study them have built reputations as experts on the basis of this. How do LGBTIQA+ people's experiences challenge these scientists' claims to expertise? What expertise do LGBTIQA+ people themselves bring to the table? When do claims of expertise oppress rather than empower? When (if ever) is it non-controversial to claim expertise in queer topics?

- **Designing community engagement processes:** STEM researchers and science communicators are increasingly asked to engage with communities of interest when designing new lines of research, technologies, and communication products. As a science communicator, how might power relations affect your interactions during such engagement, especially when engaging with diverse LGBTIQA+ communities? How can you transfer power to the individuals and communities you engage with? How can you ensure you do not take power from them?

Activities:

- **Discourse analysis:** ask your students to analyse a science communication or STEM text relevant to LGBTIQA+ people or issues. This could be a journal article, media release, video, activist campaign material, or other form of work. Whose perspectives are implicitly and explicitly outlined in the text? Whose perspectives are excluded? Does the text otherize LGBTIQA+ people and, if so, how? How might the text be improved in this regard?
- **Creative response:** encourage your students to create an artistic commentary on the relationship between science and LGBTIQA+ people and issues. It could take the form of a visual artwork, video or audio piece, creative writing, or another artistic form. Discuss the artworks as a group, sharing artist intentions and audience interpretations to reach a common understanding. Explore the pros and cons of artworks as communication media for exploring these topics compared to more conventional forms of science communication.
- **Technology scenario:** have your students work in groups to explore the meanings and potentialities of new technologies for LGBTIQA+ people. Student groups should select a new technology such as a game, a device, or an application of interest to them. They should then, through research and group discussion, determine how it might exclude or include LGBTIQA+ people and how it might serve LGBTIQA+ people's interests or be detrimental to them, and present their thoughts to the class. Then have students research technologies created by and/or for LGBTIQA+ people. Research the development process, for example any community engagement that fed into it, and anything the creators have said about their motivations. What can we learn from their experiences about inclusive technological development and innovation processes?

PART II

Representations of Queerness in Public Science Communication

4

Queering Science Museums, Science Centres, and Other Public Science Institutions

Eleanor S. Armstrong and Simon J. Lock

Queering the scene at the science institution

While approaches to queer interrogation/integration within arts and socio-historical museums are well documented (for example, Sullivan and Middleton, 2019), there is little yet written on the value of extending this approach to public science institutions. In this chapter, we think about queering public science institutions. Here we argue that that queer/ing can (and should) be an equally valuable lens for science, technology, engineering, mathematics, and medical (STEMM) institutions. We use the phrase 'public science institutions' to describe a myriad of locations where everyday STEMM learning is performed and received publicly through science communication. We also recognize that many of these public science institutions are active sites of scientific research too (for example, Singapore Botanical Gardens, American Museum of Natural History); and as such they are not only representing science but also creating the very knowledge they represent. Thus, we collate here some ideas and directions spanning zoos, botanical gardens, natural parks, farmyards, makerspaces, aquaria, open air science sites, as well as science museums and science centres, that are aimed to direct the reader to ways of engaging queer theory in science communication practices and institutions in all these contexts. However, in writing this chapter we acknowledge that any queer work can never be definitive: thus we encourage unpicking and redeveloping these in any future queer interventions into science institutions.

Public science institutions have multiple valences for publics as sites of leisure, sites of learning, or sites of (un)official pedagogy. Underpinning these, such institutions are also part of projects to create and uphold the social and structural norms of societies they claim to represent. This can be seen in existing science institution theorizing that focuses on who is in the audience at the institution, highlights who is not 'learning' to be part of the society, and interrogates what norms are being communicated as knowledge to learn (Dawson, 2019). As per other scholars (for example, Cassidy et al, 2016), we thus recognize the political implications of 'rendering things up to be viewed – [as] a key means of apprehending and "colonizing" reality' (Macdonald, 1998, p 10) in the process of display and representation of (scientific) knowledge. Indeed, representing 'facts' or 'truths' about knowledge in displays is itself a form of political governance, in turn shaping how publics understand, act, and permit themselves to act.

Thus, science communication as a process of sharing (scientific) knowledge is not separated from the power of creating (social and scientific) norms. A critical queer lens might draw attention to the colonial origin of public science institutions, where the politics of display were inherently tied in with politics of nationalism, the nation-state, reproductive obligations, and imperial fantasies of control over the world. Public science institutions were and continue to be intimately tied up with the production and maintenance of racialized (for example, Das and Lowe, 2018); classed (Wong, 2012), ableist (for example, Rieger, 2016), and gendered (Dawson et al, 2020) hierarchies and norms. Knowledge that maintains the perceived moral deviancy of queer people, normalizing binary gendering and heterosexuality (which were and are tools of empire), is reinforced through places of scientific education such as public science institutions.

In doing such queer work it is also crucial that we recognize our positionality. As white scholars from the Global North[1] we are attentive to the ways that 'queer' can operate as a neocolonial knowledge project. Queering science communication through postcolonial practices should allow, encourage, and support the circulation and normalization of pluralized genders and sexualities.[2] In the rest of this chapter we extend Macdonald's (1998) analysis of the politics of display in museums as a framework for understanding the potential for queering public science institutions. We consider both the frontstage (for example, the final exhibit as a text to be analysed, the physical infrastructures of the sites, the publics' encounters with knowledge in these sites) alongside the backstage (for example, the production processes which go into building the exhibition), while suggesting that we might extend the latter to consider the broader politics of such institutional contexts.

Queering the frontstage: what the public sees

Queer practices in public science institutions might first consider the space that publics are being invited into. We might, for example, reject the assumed gendered, cis-heterosexual physical infrastructures, pluralizing what constitutes a 'family' for family admission policies and its costings that are predicated on an implicitly heterosexual nuclear family (Robert, 2014); or using gender inclusive language ('kids' instead of 'boys and girls') to describe the imagined visitor or activity participant. Visual representations might include flying an LGBTIQA+ community or Pride flag on the building itself, on social media, or allowing employees to wear rainbow lanyards or pronoun pins. Changes to the materiality of the buildings are also possible, including building gender neutral toilets (for example Mulvey, 2019 – where these infrastructures are both oriented to disability and queer inclusive practices), or using queer inclusive signage for areas around the building such as changing rooms and lifts that doesn't feature symbols for people but rather 'puts focus on the facility not the user' (Middleton, 2021; see also Leitch et al, 2016). Such work communicates queer inclusivity publicly, which might materially facilitate queer publics attending such institutions.

If we move from the space to consider what is on display, we find queer 'issues' are very often localized to medical displays. These often centre on medicalized interpretations that historicize HIV/AIDS (Parry and Schalkwijk, 2020), which incorrectly constructs queerness (and more often, specifically cis-male gayness) as inherently tied to illness, deviance, and medical regulation (Sequeira, 2020). Otherwise, queer identity is often tied to repression or the oppression by the law, as seen in the 2013 'Codebreaker' exhibition, which discussed Alan Turing's suicide. Such norms around representations of gender or sexuality extend beyond discussions about science and scientists into broader structural issues of knowledge. Animal displays rely heavily on representations of 'natural' animal behaviours, which are usually heavily gendered and heteronormative (Machin, 2008). Anthropomorphized displays obscure the socially constructed gendered and sexual norms being communicated. Even when explicit attempts to disrupt these norms are made they can also remain flawed, reproducing homonormative ideals rather than queering gender and sexual binaries (Cassidy et al, 2016).

A range of examples demonstrates how visitors use content in institutions to make sense of their own lives and identities. This can be seen in a number of ways. Hird (2004) demonstrates how overlaying of social gendered norms onto natural history collections – in their example, of birds – obfuscates knowledge of queer behaviour that these animals do exhibit to visitors. In zoos, research by Garner and Grazin (2016) has shown that visitors describe animal behaviours that they see using the language of heteronormative family relationships regardless of the actual nature of the animal behaviours they are observing. When communicating science, rejecting the gender binary and the heterosexualization of natural and scientific worlds is a vital queer practice. For example, The Field Museum has explicitly embraced SUE (a dinosaur skeleton) as a non-binary t-rex in their public communications. This is not only instructional about queer identities for visiting publics, but has the potential to educate incoming workers in the institution too (Fleenor, 2018).

This is also seen in the ways visitors engage in institutions. Gendered behaviours are often encouraged by visiting publics – for example by better supporting boys in learning science than girls even within the same family (Crowley et al, 2001). To counter this, we encourage interventions that highlight queer (re)interpretations and absences in these institutions. These include guided tours, drag science demonstrations, queer family events, or LGBTIQA+ 'lates' (for an example of these practices, see Practice Spotlight: Queer by Nature). These examples allow publics to co-create scientific and queer identities; and to see queer as being something that both relates to science and themselves.

Queer/ing 'frontstage' practice can go further than including queer content to consider the surrounding infrastructure of display and the commitment to queer approaches. How are labels written and by/for whom? Where are these objects displayed within the structure of the gallery or the museum as a whole? Is discussion about queer identities included in the temporary display, but not in the permanent display? In temporary displays, how are the queer expressions or potentials of the content reconfigured for local audiences as they tour nation-states where cultural norms and acceptability of discussing and representing queerness varies (see, for example, Cassidy et al, 2016)? How have interactive exhibits been conceptualized along feminine/masculine learning patterns, to support learners that are not usually included in these museums (Dancstep and Sindorf, 2018)?

In thinking about 'selling' science in the shop and elsewhere we not only contend with the politics of power and knowledge, but also the politics of capitalism and reproduction. Frequently, these shops are hyper-gendered – with sections that focus on 'science for girls' (often codified by pink and sparkles) and 'proper' science toys for boys (see Onion, 2016). Inclusion of queer-friendly items for sale – for example books like *And Tango Makes Three* (Parnell et al, 2005) – both find favour with and aim to extract capital from queer inclusive visitors. The museum can also sell itself as a 'queer-friendly' venue. Queer advertising, such as ZSL London's 'Some penguins are gay. Get over it.' (ZSL, 2019), draws on visual tropes of LGBTIQA+ activism to project a queer inclusive environment through museum content. Including visibly queer couples or family units in advertising for a public science institution is rare however (for example, California Academy of Sciences, 2021).

In the 21st century, the experience of the public science institution runs beyond the physical location of the institution. Thus, digital content – social media, websites, podcasts, traditional media coverage, advertisements – are also an important 'frontstage' dimension for queer participation and science communication. Are queer and gender diverse stories highlighted by the institution? Are attempts at Pride month inclusion more than a change to a rainbow logo? Participation in hashtags, demonstrating accountability for intersectionally queer commitments, or promoting queer content for more than a single month of the year, are part of demonstrating public science institutions' commitment to queer approaches as central to their mission rather than a cynical attempt at virtue signalling or segmented audience inclusion.

Our selection of examples which explore how public institutions have engaged and unpacked particular dimensions of the science they display demonstrate how this allows them to trouble norms that are upheld within the institution, and to retell queer narratives and imaginaries in displays. The paucity of examples suggests to us, however, that public science institutions face greater resistance/reticence to engage in queer (and other identity-oriented) interventions than commensurate arts or socio-historical institutions where we see much more work being documented. Lingering notions of the need for scientific 'objectivity' remain (Cassidy et al, 2016), as does a public image of science that sees identity and by extension, relating social norms, as unimportant to the research work in science. Thus, institutions continue to reify white cis-male-heterosexual observers as the most 'objective' researchers without concern (Prescod-Weinstein, 2020). Public science institutions continue to resist acknowledgment of the material impacts of social norms on the identities of queer scientists by. See for example, recent protests around naming large scientific instruments such as telescopes after homophobic leaders (for example, Prescod-Weinstein et al, 2021) or calls for eugenists' names from buildings (for example, UCL, 2020). Pushback against such moves continues to perpetuate an artificial separation between the knowledge produced in science, and the social contexts and views of the scientists that produce it. Similarly, efforts to decolonize science museums have called for the historical, social, and cultural context of white settler colonialism to be told as part of the narratives of how scientific knowledge was, and is, shaped. A queer approach here might depart from including queer scientists, to unpick the cultural and historical context of, for example, the medicalization of homosexuality in the 19th century, or the complicity of physical scientists in developing and enforcing homophobic cultures in scientific and state workplaces.

Queering backstage: behind the scenes

Recent attempts within public science institutions towards the inclusion of queer publics, and/or the reinterpretation of their collections, frequently overlook deeper structural change within the organization. As a result, these can be dangerously tokenistic or solely driven by audience engagement and marketing concerns (Liddiard, 2004). A more radically queer approach calls for us to not only think about the content and publics but also the institutions themselves. In this section we consider what it might mean to queer some of the other dimensions that are less 'public' in public science institutions, organized along three dimensions – people, funding, and the constituent parts of the institution.

People are fundamental to the functioning of any institution. Across the science museum sector there is not much existing data on LGBTIQA+ employment. Sector-wide LGBTIQA+ demographics and institutional organizational documentation (for example workforce surveys) have the potential to illuminate institutional gaps in support and employment of LGBTIQA+ folk. In addition to this, better organizational support is required within structures of the institution. This includes engaging with community networks, or organizing events for Pride, which should be financially supported rather than institutions expecting this as additional free work from already marginalized staff. Further, inclusive physical infrastructures and staff policies for queer staff (for example equal parental and medical care leave, insurance cover, and navigable instructions for name changes) are also vital for those employed at institutions (for example, Vincent, 2016).

Institutional support of queer employees must pay attention to the uneven politics of safety and inclusion within the LGBTIQA+ communities where queer people of colour and trans communities are less well supported (LGBTQ+@cam, 2019). The Exploratorium's 'Striving for Trans-inclusion and Anti-Racism in Science learning' (STARS) programme specifically employed trans and queer youth to undertake research and education on inequities foregrounding their perspectives and understanding of community needs (Exploratorium, 2021). Who (if anyone) within the LGBTIQA+ community occupies positions of responsibility and decision-making within the organization is also key. LGBTIQA+ employees are underrepresented in decision-making positions (such as at the board level) in arts and sociocultural museums (Arts Council England, 2016) and we anticipate this to be the case in public science institutions too. Active learning within (and between) communities is also important. Platforming work at international forums, supporting networks of LGBTIQA+ employees between institutions, and developing networks of practice can support emotive and collegiate growth and learning within and between similar contexts.

Public science institutions are costly to run. Funding primarily comes through either state funding and/or private or corporate donors. Both streams may present limitations on queer possibilities for the institution – state regulation of what the money can be spent on (Adams, 2021) or state-level regulation of displaying and discussing queer identities can limit what is deemed appropriate for 'the public' and how narratives within these public institutions can be constructed or challenged through a queer science communication. We know too that private or corporate donors can censor the materials that are permitted on display (Culture Unstained, 2021). Awareness of both of these tensions may additionally lead to self-censoring on the part of the employees of the institution.

Finally, we turn to the collections that underpin these public science institutions and how they might also be developed with a queer orientation. We see many possibilities for queering practices in these institutions. Reinterpreting existing collection displays (Wellcome Collection and Research Centre for Museums and Galleries (RCMG), 2020) and updating research practices (Leitch et al, 2016) can work to improve existing records that are publicly accessible as well as the material content of the collection. When considering digital collections, Clewlow (2021) has demonstrated how different terms bias searching within online archives, showing that terms such as 'couples' returned primarily heterosexual results, whereas 'gay' returned mostly results to do with HIV/AIDS in The Science Museum Group's digital records. These biases shape the perception of what is even available within the archive to use in displays, as well as directly shaping normative understandings of what these keyword terms mean. Critical reflection on the process of collecting objects, creating records, and digitizing collections, focuses in on who has the power to do this, and how knowledge is produced about queer people and the world.

Conclusion

We feel there is, in general, significantly greater interest, activism, and research on what we have described as 'frontstage' queer practices than there is on the 'backstage' activities at public science institutions. While we celebrate all the work that is being done to queer public science institutions globally, we also call for further research and practices across the board. In contrast to both arts and socio-historic museums, STEMM institutions lack the depth and breadth of research. Public science institutions have also thus far fallen out of the literature on broader queer-oriented institutional critiques, including those on academia or STEMM fields; which may also be reasons that activities in public science institutions are less well theorized.

We urge researchers to resist separating 'frontstage' practices of science communication from the 'backstage' institutional context in which it is produced. Our authorial sleight of hand to set up this chapter does not reflect a material difference: it is the queer people employed or volunteering in institutions or activists lobbying for change from outside that push forward public-facing changes. We have extended Macdonald's theorizing of the political tensions that run through public science institutions to explicitly focus on the entanglements of queer politics in institutional contexts that shape and constrain their potentials. We have used queer as a method to direct our disruption of the sanctification of particular representations of knowledge that support the colonial, capitalist, white supremacist, heteronormative, gendered ideas that are both created by science and then reproduced in communication by public science institutions. We reiterate the tensions of assimilatory practices that might attempt to subsume queer identities under the colonial auspices of the public science institution via the flying of a Pride flag ('pinkwashing'), and instead direct the reader to the potentials of radical structural changes that could take place within the institution via queer action.

Queering the practices of the public science institution can disrupt the power of the knowledge structures the institution itself is intended to uphold. If, as we argued at the start of this chapter, these public science institutions are mechanisms for educating and inducting publics into conforming to the (neo)colonial norms through scientific logics, what would rejecting these norms through queer approaches look like in practice? Wayne Modest (Taking Care, 2019) argues that public institutions have the potential to radically reimagine what 'caring' for artefacts might look like. What if, instead of thinking about 'care of the collection', institutions were 'caring and careful spaces'? Through this reorientation towards care for communities, individuals, histories, heritage, or memories, there is the possibility to reject often violent acts of categorization, within colonial, cis-heteronormative disciplining of scientific knowledge systems. What would the queer science institution look like if we, as encouraged by Muñoz (2009), rejected the idea that these archives were complete and represented 'truths', and instead fundamentally reimagined the institution outside of colonial practices? Might we instead accept multiple readings of 'science', 'technology', 'engineering', 'mathematics', and 'medicine'? Queering science communication in the context of public science institutions invites us to be provocative about what constitutes 'science', who practices it, and how public science institutions are operationalized as educational tools of the state.

Notes

[1] SJL uses he/they pronouns and is a queer white British academic in their 40s, they are genderqueer and polyamorous. ESA uses she/her pronouns and is a queer, dyslexic, white British academic (living in Sweden) in her 20s.

[2] In putting this chapter together, we are particularly aware that there is a Euro-American-centric bias in the existing literature on this subject, where the same bias is not as marked in other areas of queer museum scholarship.

References

Adams, G.K. (2021) '"Polite but managed" summit fails to allay concerns about government overreach', *Museums Association* [website] 26 February, Available from: https://www.museumsassociation.org/museums-journal/news/2021/02/polite-but-managed-summit-fails-to-allay-concerns-about-government-overreach/# [Accessed 5 August 2021].

Arts Council England. (2016) *Equality, Diversity and the Creative Case*, London: Arts Council England.

California Academy of Sciences. (2021) 'New exhibit opening June 24 highlights LGBTQ+ intersectional identities in STEMM', *California Academy of Sciences* [media release] 22 June, Available from: https://www.calacademy.org/press/releases/new-exhibit-opening-june-24-highlights-lgbtq-intersectional-identities-in-stemm [Accessed 5 August 2021].

Cassidy, A., Lock, S.J., and Voss, G. (2016) 'Sexual nature? (Re) presenting sexuality and science in the museum', *Science as Culture*, 25(2): 214–238.

Clewlow, M. (2021) 'Preserving queerness of community LGBTQ+ archives', in M. Bühner, R. Rinner, T. Tammer, and K Töpfer (eds) *Sexualitäten Sammeln: Ansprüche und Widersprüche im Museum*, Böhlau Verlag, pp 101–112.

Crowley, K., Callanan, M.A., Tenenbaum, H.R., and Allen, E. (2001) 'Parents explain more often to boys than to girls during shared scientific thinking', *Psychological Science*, 12(3): 258–261.

Culture Unsustained. (2021) 'Science Museum Group's partnerships with Adani, Shell, BP & Equinor', *Culture Unstained* [website], Available from: https://cultureunstained.org/sciencemuseum/ [Accessed 5 August 2021].

Dancstep, T. and Sindorf, L. (2018) 'Creating a female-responsive design framework for STEM exhibits', *Curator: The Museum Journal*, 61(3): 469–484.

Das, S. and Lowe, M. (2018) 'Nature read in black and white: decolonial approaches to interpreting natural history collections', *Journal of Natural Science Collections*, 6: 4–14.

Dawson, E. (2019) *Equity, Exclusion and Everyday Science Learning: The Experiences of Minoritised Groups*, London: Routledge.

Dawson, E., Archer, L., Seakins, A., Godec, S., DeWitt, J., King, H., Mau, A., and Nomikou, E. (2020) 'Selfies at the science museum: exploring girls' identity performances in a science learning space', *Gender and Education*, 32(5): 664–681.

Exploratorium. (2021) 'STARS at the Exploratorium', *Exploratorium* [website], Available from: https://www.exploratorium.edu/education/professional-development-programs/stars [Accessed 5 August 2021].

Fleenor, S.E. (2018) 'How a T. Rex named SUE became a nonbinary icon', *Them* [online] 4 May, Available from: https://www.them.us/story/sue-the-t-rex-is-a-nonbinary-icon [Accessed 5 August 2021].

Garner, B. and Grazian, D. (2016) 'Naturalizing gender through childhood socialization messages in a zoo', *Social Psychology Quarterly*, 79(3): 181–198.

Hird, M.J. (2004) 'Naturally queer', *Feminist Theory*, 5(1): 85–89.

Leitch, C., Youngs, R., Gavigan, A., Lesperance, M., Burns, W.J., Cohen-Stratyner, B., and Hansen, J. (2016) 'LGBTQ Welcoming Guidelines for Museums: developing a resource for the museum field', *Museums & Social Issues*, 11(2): 138–146.

LGBTQ+@cam. (2019) *Out at Cambridge: Why LGBTQ+ disclosure matters to individuals and institutions*, Department of Sociology, University of Cambridge.

Liddiard, R. (2004) 'Changing histories: museums, sexuality and the future of the past', *Museum and Society*, 2(1): 15–29.

Macdonald, S. (1998) 'Exhibitions of power and powers of exhibition', in S. Macdonald (ed) *The Politics of Display: Museums, Science, Culture*, London: Routledge, pp 1–24.

Machin, R. (2008) 'Gender representation in the natural history galleries at the Manchester Museum', *Museum and Society*, 6(1): 54–67.

Middleton, M. (2021) 'As a designer who consults with museums on gender inclusive practice, I am often asked about signage and icons. Here are some of my top recs for easy swaps you can make for a more gender inclusive visitor experience:' (untitled Twitter thread), *@magmidd* [Twitter account] 2 July, Available from: https://twitter.com/magmidd/status/1410909224490700801?s=20&t=O3F2o1LCXiXSCWz6AYzqEA [Accessed 28 March 2022].

Mulvey, M. (2019) 'Going public(ly): what can public responses to changing toilet provision in cultural institutions reveal about attachments to gendered spaces?', Paper presented at *Stalled! Beyond gender-segregated public toilets and towards equitable design*, 9 March 2019, UCL.

Muñoz, J.E. (2019) *Cruising Utopia*, New York: New York University Press.

Onion, R. (2016) *Innocent Experiments: Childhood and the Culture of Popular Science in the United States*, Chapel Hill: The University of North Carolina Press.

Parnell, P., Richardson, J., and Cole, H. (2005) *And Tango Makes Three*, New York: Simon & Schuster Children's Publishing.

Parry, M.S. and Schalkwijk, H. (2020) 'Lost objects and missing histories: HIV/AIDS in the Netherlands', in J.G. Adair and A.K. Levin (eds) *Museums, sexuality, and gender activism*, London: Routledge, pp 113–125.

Prescod-Weinstein, C. (2020) 'Making Black women scientists under white empiricism: the racialization of epistemology in physics', *Signs: Journal of Women in Culture and Society*, 45(2): 421–447.

Prescod-Weinstein, C., Tuttle, S., Walkowicz, L., and Nord, B (2021) 'The James Webb Space Telescope needs to be renamed', *Scientific American* [online] 1 March, Available from: https://www.scientificamerican.com/article/nasa-needs-to-rename-the-james-webb-space-telescope/ [Accessed 5 August 2021].

Rieger, J.L. (2016) *Doing Dis/Ordered Mapping/S: Embodying Disability in the Museum Environment*, PhD Thesis, University of Alberta.

Robert, N. (2014) 'Getting intersectional in museums', *Museums & Social Issues*, 9(1): 24–33.

Sequeira, R. (2020) 'Death of a museum foretold? On sexual display in the time of AIDS in India', in J.G. Adair and A.K. Levin (eds) *Museums, Sexuality, and Gender Activism*, London: Routledge, pp 101–112.

Sullivan, N. and Middleton, C. (2019) *Queering the Museum*, London: Routledge.

Taking Care Project (2019) 'Taking care – ethnographic and world cultures museums as spaces of care', *Taking Care* [website], Available from: https://takingcareproject.eu/about [Accessed 5 August 2021].

UCL. (2020) 'UCL denames buildings named after eugenicists', *UCL News* [online] 19 June, Available from: https://www.ucl.ac.uk/news/2020/jun/ucl-denames-buildings-named-after-eugenicists [Accessed 5 August 2021]

Vincent, J. (2016) *LGBT People and the UK Cultural Sector: The Response of Libraries, Museums, Archives and Heritage Since 1950*. London: Routledge.

Wellcome Collection and RCMG. (2020) *An ethical approach to interpreting disability and difference*, Research Centre for Museums and Galleries, Available from: https://wellcomecollection.org/works/jywe7h9f [Accessed 28 March 2022].

Wong, B. (2012) 'Identifying with science: a case study of two 13-year-old "high achieving working class" British Asian girls', *International Journal of Science Education*, 34(1): 43–65.

ZSL. (2019) 'Some penguins are gay. Get over it', *ZSL London Zoo* [website] 26 June, Available from: https://www.zsl.org/zsl-london-zoo/news/some-penguins-are-gay-get-over-it [Accessed 5 August 2021].

Practice Spotlight: Queer by Nature: The LGBTQ+ Natural History Tour

Josh Davis

When it comes to science museums – and natural history museums in particular – many have flown under the radar when it comes to addressing subjects such colonization and LGBTIQA+ themes. But queering natural history museums can offer several unique perspectives.

Like other museums and galleries, there are the stories of the LGBTIQA+ people behind the collections such as the curators, scientists, collectors, and artists. But in contrast to many of these spaces, there are also queer stories within the specimens themselves as the animals and plants were once living organisms, expressing their own queer behaviours.

These can allow for an interesting and often surprising delve into just how diverse the natural world is when it comes to both animal behaviour and how many species organize their sex (and, depending on definitions, gender). From species of lizard that are all female to fish that can produce both sperm and eggs before self-fertilizing, the natural world really does break all the (perceived) rules.

Exploring the deeply queer side of nature can offer insights and expose our inherent cis-het biases and assumptions about what we think about sex, sexuality, and gender. This manifests itself in many different aspects including, but not limited to, the language we use for what we consider 'normal'.

The Natural History Museum, London (NHM) developed and ran its first LGBTQ+ Natural History Tour in June 2019 for Pride Month. The intention was to test out ideas for a more regular LGBTIQA+ themed tour for the museum. As the aim of the tour was to reach diverse audiences not usually served by the museum, members from the Digital Media, Interpretation and Science teams worked together to produce a tour that highlights stories and features content that celebrates the full diversity of nature while making visitors think about how they look at the natural world, and how their own biases might colour their interpretations of it.

To begin with, staff from the Digital Media team took various objects and discussed how they link to either LGBTIQA+ themes, people, or history. Because this was new content created specifically for the tour, these pieces of text were then shared with and further developed by both the Interpretation and Science teams to make sure everything was accessible and accurate.

After having developed a basic narrative for the tour, the NHM team then carried out a workshop in collaboration with the LGBTIQA+ community. As this project was very much a collaborative piece of work among LGBTIQA+ staff within the museum, it was important for all those involved to open this discussion out to the wider community.

The objects and their stories were presented to members of the LGBTIQA+ community who were invited to the NHM to help decide on the tour's final content. This process involved open discussion and critique of the objects selected, the themes explored, and the language that should be used.

Eventually, it was decided to keep the tour contained to the main hall of the museum, with themes centring on how historic and current societal biases are imposed on science and the natural world and the implications this has for the interpretation of animals and their behaviour.

But by using what is on display as a jumping off point, it was possible to create an LGBTQ+ Natural History Tour that was fun, engaging, educational, and on theme. For example, it was possible to use the large piece of *Turbinaria* coral on display to open up a discussion about how one third of all coral reef fish are hermaphroditic, the *Mantellisaurus* fossil to talk not only about the sex bias in dinosaur names but also the queer palaeontologist Franz Nopcsa, and the statue of Charles Darwin to delve into the shortfalls and heteronormativity of his theory of sexual selection.

The tour was then run for the first time during Pride Month as a free, ticketed event with each tour limited to 20 people each and running to 40 minutes long. At the end of each tour feedback was gathered, showing that it was incredibly well received with 93% of respondents saying they would recommend the tour to a friend.

Asked 'what was something interesting that you learned at the tour?', one of the visitors replied: "Just how far back the knowledge/awareness of [LGBTIQA+ behaviour] goes, and how the suppression of information has affected the knowledge and understanding we have today."

Following on from this, the LGBTQ+ Natural History Tour was further developed as a self-guided tour for Valentine's night and once again for the NHM's YouTube channel. It is now currently being adapted by the NHM Events team for a semi-regular tour on offer to visitors.

Practice Spotlight: Science Queers: Overacted Representation in Science Communication

Òscar Aznar-Alemany

First steps

Science Queers is a grassroots project based in Barcelona with the double mission to promote queer identities and science, led by the Catalan drag queen scientist Lana Vuli.

Our first performance was in January 2017 at the final conference of the project ECsafeSEAFOOD. Lana's makeup skills were still rough, but the Seafood Tango, a drag burlesque act about seafood safety, proved that entertainment and queer identities did not hinder the rigour or accuracy of the science presented and that they were welcomed and applauded. Months after our ten-minute musical, attendees still talked about it and were able to repeat some key information (and lyrics), which never happened with its traditional presentation version.

For two years, the Seafood Tango and another act toured in conferences, bars, and even the Spanish Nowhere event, where strangers would approach us at 2 am on the dance floor with questions about science. Most of those events had one thing in common: they were organized by well-established non-queer entities outside an LGBTIQA+ context. We were exactly where we needed to be.

Science Queers is a much-needed exercise of overacted representation for closeted scientists and non-queer audiences. Making queerness a key element of science communication results in people publicly showing their support for our identities with their reactions, which they might not do when not prompted. This, in turn, empowers LGBTIQA+ people to be themselves in public and in the laboratories.

Exploring formats

In 2019, our YouTube channel was launched as Lana's one-woman show. However, uploading videos is less effective than taking the act to physical spaces, especially in a platform already full of both drag queens and science communicators. Therefore, Science Queers Academy (SQA), a live and video hybrid event, was created.

SQA is a reality show in which drag artists communicate science. The performances of SQA are filmed with a live audience in non-queer venues. Live events are more enjoyable and allow us to deliver our product to a target audience. Performing in non-queer venues means actually taking representation to society outside of our safe spaces. Restricting the project to LGBTIQA+ venues would make it just queer entertainment.

In the two months when the first season of SQA was published, our YouTube subscribers and video views skyrocketed. But were we reaching the right audience?

For LGBTQ+ STEM Day 2020 and 2021, we held other live events including short talks or shows by non-drag members of our community who work in STEM. While the events sold out, the audience was mostly queer. We were delivering science to queer people and growing popular in the queer circles. Some drag queens in Barcelona had been heard to chant 'Science Queers!' when something clever was said and we were becoming a trusted source to answer our friends' scientific questions. However, we were neglecting the objective of taking queer representation to non-queer crowds.

Looking forward

Our project was young, lacking resources, and perhaps too queer and too niche to grow fast. Nevertheless, in the two years before YouTube, our appearances proved extremely popular among cis-straight audiences; we won awards and people still congratulated us on the shows even months after them. This suggests that not only is society ready for LGBTIQA+ representation (and the campiest deliveries), but our STEM peers and the general public acknowledge the need for it and the positive impact it has on promoting inclusivity.

However, after sharing a queer science carol in a science communication mailing list, we received a reply from a (cis-straight male) retired university professor stating that the carol was a mistake and wondering what a sex therapist would think of it. The moderators of the list argued that the man had no ill intent and blamed it on a misinterpretation due to the lack of context or voice intonation of an email. They did take offence at our calling out their bias, though.

This incident was a reminder of how much work was left to do and how little influence we still had. It was necessary to take Science Queers back to its origins and participate in external events instead of focusing on our own. Additionally, it was a reminder that when we did plan our own events, we should collaborate with more influential and non-queer associations or institutions.

That's how 2022 became our outreach year. Collaborating with well-established organizations allows Science Queers to organize events in research facilities showcasing their own queer staff, to make academia aware of and familiar with us, thus considering us in all scenarios, and to include queer representation in other science communication actions directed to the general public.

In a nutshell, STEM is ready for LGBTIQA+ identities and, if STEM includes and embraces us, the chance is that the rest of our society will follow.

Practice Spotlight: Science is a Drag! Online Events

Carla Suciu, Brynley Pearlstone, and Sam Langford

Join us in Glasgow, Scotland, for a moment, travelling back in time to the beginning of the dreaded year 2020. The entire country has been placed on lockdown, much like others globally, and three LGBTIQA+ people working in STEM are at a loss – their plans to host a science cabaret focusing on combining science with the art of drag for the local science festival at a standstill. Then, comes an idea: 'Well, why don't we just do it online?' It was a gamble, and it could have flopped entirely.

However, by tapping into their under-served community and bringing LGBTIQA+ events that LGBTIQA+ people themselves created, Science is a Drag took its online stage. The first show itself brought in 180 tickets, generating a profit of over £1,000 with a budget of zero. The social media response (monitored on Twitter using #ScienceisaDrag and through the comment section on YouTube) was overwhelmingly positive. Financially, for the performers, Science is a Drag proved lucrative. We are passionate about ensuring fair pay for our performers. This meant each act and host on the show received a share of over £100 each for their efforts, a figure higher than the standard rate for a drag event. By splitting the ticket sales between the hosts and performers, Science is a Drag has been able to pay well above the industry standards consistently – often double. This was crucial when COVID-19 shut down queer performance spaces, leaving no appropriate substitute behind.

Back at headquarters, the organizers gather as they do to this day after each show to discuss the things that did not go quite so well, or feedback received. As the first show was delivered with zero budget, using free tech solutions, this resulted in glitchy live sound from the hosts and live guests, plus slow transitions between pre-recorded acts and live interviews.

For the next date, just before LGBTQ+ STEM day, the team decided to take a risk based on the positive reaction of the first event. It was decided it was worth the gamble of investing in third-party tools – this included

Streamyard to give a professional look and feel to the show as well as live captioning via Otter.ai.

Promising to resolve their technical issues and put on an even smoother night, sales for the second show outstripped the first. For evaluation, a follow-up survey was circulated to attendees, asking about the enjoyability of the event, value for money, and general feedback. Enjoyability and value for money ranked highly, alongside the format and the accessibility features employed. Audiences indicated that they were keen to see even more in-depth scientific content as part of the performances.

Because of this, the show has gone from strength to strength, with rebranded promotional materials and a custom logo designed by a former cast member. The next show drew the attention of Penn State University, which bought $400 worth of tickets for their student body, as well as the magazine *We Are Europe*, which highlighted the event in their Pride Month issue in June 2021. With almost 300 tickets sold for this show alone, the little cabaret night had outstripped the organizers' initial imaginations.

The show has now featured globally sourced acts and an audience from four continents showing even further growth.

From that anxious and turbulent time at the beginning of the pandemic, Science is a Drag has gone on to create a name for itself in queer science communication. It has given LGBTQ+ researchers and performers a space to perform science.

Will the return of in-person events mean the end of Science is a Drag? Absolutely not. As the world attempts to bring about a new normal, so too will Science is a Drag! What will the future look like as the world moves to hybrid style events? The organizers hope to combine the access of an online show with the glitz and drama of an in-person event.

One thing is for sure, there is a clear appetite for science communication through a queer lens. Science can be a drag, but it doesn't have to be!

5

Queer Characters in Science-themed Fiction

Lindy A. Orthia and Leo P. Visser

Media representations of scientists have been a mainstay topic of science communication research for decades.

In addition to examining representations of scientists and other STEM professionals such as engineers and mathematicians in news media, science communicators have examined their representation in fiction texts such as films, television, theatre, novels, and comics. They have conducted this research for a range of reasons including to evaluate equality of representation among scientist characters by genders, ethnicities, and disciplines (for example, Long et al, 2010), and to investigate how characters in science-themed fiction promote particular ideologies such as scientism (for example, Orthia, 2011).

The ultimate rationale is often to understand what impact fictional representations of STEM professionals have on public discourse, for example on people's concepts of what scientists are like or who can be an engineer. Sometimes this is framed as investigations of fiction's role-modelling capacity for STEM careers (for example, Steinke et al, 2012). In other cases, the rationale is to sample fiction as a manifestation of existing public discourse, to find out what a particular society already thinks about science, its personnel, and the 'others' it contrasts itself to (for example, Haynes, 1994).

However, almost none of this work has attended to fictional representations of queerness. That is the case whether it refers to diversity in scientific characters' sexual orientation, gender, and sex characteristics, or fiction's depictions of queer issues in science.

Indeed, diversity in gender and sex characteristics has been rendered still more invisible than usual by the field's analytical conventions. Most studies examining the gender or sex of scientist characters look only at differences between 'women' and 'men', implicitly assuming a cis-binary model of gender and sex that is not questioned or even mentioned. Such assumptions patently exclude gender diverse people and people with diverse sex characteristics who don't identify as female or male. Non-heterosexuals are also often overlooked by analysts who presume characters are heterosexual unless canonically proven otherwise. As well as perpetuating harm to queer people specifically, such analytical decisions do the field a general disservice by keeping discourse about inclusion at a superficial level.

So what might a science communication analysis of queer representation in science-themed fiction look like? And, given science communication is generally an audience-oriented field, how do we account for queer audience interpretations of science-themed fiction? This chapter outlines some key issues in this space in the dual hope that future science communication researchers can build on them and that science communication practitioners who create and reflect on fiction can respond to them.

We approach the topic in two halves. We first explore queer science communication approaches to fiction analysis using examples of canonically queer characters in two recent television programmes. We then discuss implications of reading queerness into STEM professional characters when the text does not explicitly label them queer.

Queerness and sciencyness in canonically queer scientist characters

The most obvious approach to studying queer representation in science-themed fiction is to analyse STEM professional characters who are canonically lesbian, gay, bisexual, trans, intersex, asexual, or otherwise queer+. For brevity hereafter, we call such characters 'queer scientists' unless more specific terms apply.

The limited research done in this area shows queer scientists in fiction can make a difference to audiences who want to see people like themselves on the screen or page. An audience study of television programme *Orphan Black* (2013–2017), which features lesbian evolutionary biologist Cosima Niehaus as a central character, showed character diversity that deepens the connection between audiences and characters can validate the career ambitions of diverse future scientists, and enhance non-scientists' appreciation for science (Smith and Tucker, 2020). Representation matters. More research in this space would be valuable.

There are potentially several ways to analyse scientist characters from a queered science communication perspective. One is to survey the kinds of queer identities they represent and whether they conform to stereotypes. Another is to analyse what sort of scientists they are, in terms of traits like scientific credibility and ethics, to identify ideologies that associate queerness with particular orientations to science. A third is to analyse queer themes in stories they appear in, whether literal or metaphorical. All can help us connect societal ideologies about queerness to societal ideologies about science and understand real world discourses that shape the multitude of ways queer people interact with science.

The general paucity of queer scientist characters is the biggest obstacle to conducting such analyses. Their representation has improved in the 21st century, but even today remains patchy. Many more scientist characters are *implied* to be queer or interpretable as queer, which we discuss later. Canonically queer scientists are harder to come by.

To illustrate our analytical approaches we examine two television programmes that break the mould because they feature more than one queer scientist: the *Doctor Who* spinoff *Torchwood* (2006–2011) and one of the newest instalments in the *Star Trek* franchise, *Star Trek: Discovery* (2017–present).

Fantastical pan/bi sciencyness in Torchwood

The science fiction series *Torchwood* has become iconic because of the loosely defined pansexuality or bisexuality of its main characters. Almost all are implied to have had sexual encounters with people of the same gender or alien creatures of unknown gender. This queerness was admittedly patchy, vague, and not always positive, and most characters were not explicitly labelled with a queer term. Some commentators have criticized the promise of sexual fluidity in *Torchwood* as a tease not a reality (Bradford, 2014), aside from its most reliably queer character, the show's pansexual male lead Captain Jack Harkness.

The main characters display proficient technical and scientific skills and knowledge commensurate with being STEM professionals. However, they are not conventional scientists because they are experts in alien life and alien technology who work for a secret security organization. Some have scientific qualifications in medicine and IT, but insofar as they represent 'science' it is as general futuristic techy-ness rather than conforming to any real-world scientific norms. Most obviously, Jack is a con artist from Earth's future, not a scientist. His technological prowess comes from his future roots, not scientific training, though it is an important part of his character.

What is the significance of this association between queerness and sciencyness in *Torchwood*? Certainly, the presence of queer characters in any television show was unusual in the 20th century, and having multiple queer characters was still rare when *Torchwood* was made. In this sense, its primary relevance to queer people was it increased the visibility of queer sexual attraction. For example, in a study of queer science fiction fans, Kerry (2019, p 110) cites a *Torchwood* fan who said, 'It is important for me to feel represented in the TV shows that I watch and these days I tend to only seek out shows that include queer people.' That benefit, while undoubtedly important in its own right, may also extend to role-modelling the existence of queer people in science-related careers. In addition, *Torchwood* links pro-science and pro-queer ideologies by creating a portrait of Earth's future as a version of liberal, Western techno-cosmopolitanism. This no doubt appeals to queer fans of Western science but may carry less comfortable implications for queer audiences opposed to the continued global hegemony of Western science or who wish to decouple queerness from Western values.

Torchwood is less progressive around gender diversity than sexual attraction. While sexually open, its main characters are not gender diverse. They are quite mainstream and binary-gendered in appearance, having this in common with other prominent queer scientist characters such as lesbian/bisexual computer geek Willow Rosenberg from *Buffy the Vampire Slayer* (1997–2003; Orthia, 2010). A cis-binary gender model is also reinforced by events in *Torchwood*'s second episode. The story concerns an alien parasite that feeds on energy produced by orgasms, so it seeks sex with humans. In a disappointing lesploitation moment, the parasite seduces then rejects sex with *Torchwood*'s female lead, claiming 'It's no good, it's got to be a man.' In addition to implying men's orgasms possess some mysterious, unique property, the scene reifies a biologically essentialist, cis-binary model of gender and sex. The sciency flavour of the programme and the alien's ostensibly objective view of human sex/gender would both seem to validate that oppressive ideology.

Sexuality and gender affirming science in Star Trek: Discovery

Like *Torchwood*, *Star Trek: Discovery* depicts a future built on Western-style technologies. But its depiction of queer scientist characters is rather different. A key distinction is it explores the specific natures of characters' scientific, sexual, and gender identities. Perhaps because it is set in humanity's future, not the present day, it uses real-world references to ground its depiction of that future, in contrast with *Torchwood* which is set in the present day and has fantastical depictions of future worlds. *Torchwood* therefore has ungrounded, vague discourse around science, sexuality, and gender, perhaps to avoid dating it, while *Discovery* seeks to acknowledge current real-world discourse to enhance its relevance. This may be a distinction future researchers wish to examine in more depth.

Star Trek: Discovery normalizes queerness in two ways. The first is its unremarked upon inclusion of non-heterosexual characters. *Discovery* features *Star Trek*'s first openly gay couple: medical officer of the eponymous starship USS Discovery, Dr Hugh Culber, and his partner, the ship's science officer and astromycologist, Lieutenant Paul Stamets. Their relationship is depicted in domestic detail and is important to the plot, yet does not detract from the characters' scientific credibility. In this it differs from the way female scientists' intimate relationships are often depicted in fiction, which tends to undermine their scientific credibility (Flicker, 2003). In season 2 of *Discovery* the ship also rescues the stranded Commander Jett Reno, former Chief Engineer of another ship. She joins the Discovery science team, and her queerness is confirmed when she discusses her wife. These characters each possess specialist scientific knowledge and skills rather than a general futuristic techy-ness. This means that, as far as queer scientist role models go, they are relatively grounded and three-dimensional, carrying on a *Star Trek* tradition of promoting STEM careers to minoritized people (Penley, 1997).

Discovery also normalizes queerness through more conscious, careful explorations of gender diversity. In season 3 the ship encounters the technologically brilliant and non-binary Adira, who also joins the science crew. While initially addressed by colleagues using she/her pronouns, Adira asserts their non-binary status and they/them pronouns are used after a few episodes. It is a significant moment for the visibility of gender diversity in popular television.

Adira is in a relationship with a character called Gray who is described in promotional materials as trans (Startrek.com, 2020). Gray physically died but Adira can still interact with him, and when visiting a planet whose whole environment is charged with holographic simulations, Culber and others can see and interact with him too. Gray is disheartened to leave the planet knowing he will once again disappear for them, but Culber reassures him that, together with Adira and Paul, he will find a way to help Gray 'be seen – be truly seen – by everyone.' While this is a science fiction plot and thus inherently ambiguous (Lavender, 2011), it can be interpreted as a metaphor for medicine, science, and engineering creating gender-affirming technologies to help trans people be seen as themselves. In this sense it provides a role model for real-world cis-binary scientists on how to engage with trans and gender diverse people.

As an aside, it is notable that all the actors in these *Discovery* roles share their characters' queer sexual and gender identities (Vary, 2020). This is a history-making development that makes queer diversity visible in new ways and ensures it is not relegated to science fiction status.

Scientifically credible queers and transforming scientific culture

The queer scientist characters in *Torchwood* and *Discovery* make an important contribution to STEM inclusion by counteracting ideologies prevalent in Western culture that associate queerness with bad science. Benshoff (1997) has discussed ways that horror fiction texts associate homosexuality with evil science, for example by depicting scientists creating life without heterosexual union. Orthia and Morgain (2016) studied hundreds of scientists in science fiction television series *Doctor Who* (1963–present) and found that whenever a storyline called for non-credible, utterly failed science, the scientist characters responsible were depicted as gender non-conforming, gender transgressive, queer, and/or disrespectful of masculinist norms. The depiction of queer scientist characters who are highly scientifically credible may trouble such prevalent ideologies.

Queer characters who bring queer ways with them into the lab, and thus challenge a cis-hetero-masculinist scientific culture, are particularly needed in science-themed fiction. Inclusion is not just a matter of adding queer personnel to the existing scientific culture; we need to transform the culture itself. Orthia and Morgain (2016) concluded *Doctor Who* supports equal opportunity in scientific jobs provided all scientists accept scientific culture's masculinist norms. It remains to be seen if *Star Trek: Discovery* – or texts like it that depict an ensemble of queer scientists working together in a scientific workplace – can undermine Western science's cultural norms and thus make genuine change.

When we don't know if a scientist character is queer

This kind of analysis is all very well for canonically queer scientist characters, and we welcome the time when there are so many that our analyses become richer and deeper.

The situation instantly becomes trickier for scientist characters whose sexuality, gender, and sex characteristics are not explicitly discussed by the text.

Characters are fictional so we don't know

A key issue for analysts to consider is the fact that a character's sexual orientation, gender, and sex characteristics cannot be known for sure unless the text makes them clear (Orthia and Morgain, 2016). Sex characteristics that mainstream Western society uses to 'determine' a person's sex or gender – genitals, chromosomes, hormones – are not visible in most fiction. So-called secondary sexual characteristics such as breast size, facial hair, and voice register are more apparent, especially in performed fiction such as films, television, and theatre, but there is wide variation in them among humans irrespective of gender and they can also be altered. What a person feels, who they are attracted to, how they see themselves, and how they define their identity are also invisible to others unless made explicit by the text. Unlike real people, we cannot ask a fictional person about their identity, so we can only work with what the authors give us. These issues are compounded when considering alien and other non-human characters in science fiction. Orthia and Morgain (2016) excluded all non-human characters from their study of *Doctor Who*'s scientists because of the difficulty of classifying them by typical Western notions of gender (how do you gender a Dalek?).

The implication is that science communicators who analyse representations of gender among scientist characters make assumptions based on stereotypes, unless the text explicitly identifies a character's gender. The same is true for analyses of characters' sexual orientation. Most assume cis-binary genders and heterosexual identities unless a fiction text states otherwise. This falsely perpetuates queer invisibility within science communication writings.

Analysts also largely ignore the possibility that a character may have variations in sex characteristics outside Western binary expectations. Such variations are rarely depicted in fiction itself either, and when they are depicted it is often to use them as a gimmick or plot twist. For example, the second season of lesbian-themed Australian crime drama *Janet King* (2014–2017) revealed the perpetrator of a murder was a person with Androgen Insensitivity Syndrome, so, in the words of the show's dialogue, 'the DNA reads male, it's XY, but because of the mutation, they'll look female.' This revelation was accompanied by a longer explanation in more scientific language, giving the impression of objectivity and value-free neutrality, yet the plot implies the intersex person used their sex characteristic variants for duplicity and deception. While the character might be seen as a welcome development for intersex visibility, its value for positive representation is questionable. The widespread availability of scientific information about variations in sex characteristics was arguably exploited in this instance and others to titillate and shock viewers rather than to affirm the diverse variation of sex characteristics inherent to human populations. Creators of science-themed fiction are in a position to change such representations. Analysts also have a role to play by avoiding assumptions about the nature of characters' sex characteristics, and acknowledging the statistical likelihood that, on average, 1.7% of all fictional characters will have sex characteristics that don't fit normative medical or social expectations for female or male bodies, as is the case in the real world (Intersex Australia, 2019).

There are unexplored implications of these assumptions about characters' gender and sex characteristics in other science communication research. For example in the famous 'draw a scientist' test, how can analysts be sure of a scientist's gender in a child's drawing? Future researchers might re-analyse a set of these drawings with a more open mind about this.

An approach that would help liberate research into character diversity in science-themed fiction would be to avoid all such assumptions and analyse characters from a starting point of nil gender, nil sexual orientation, and nil sex characteristics. Doing so may reveal how often fiction explicitly identifies scientist characters' gender and sexual traits, and how often it merely alludes to such matters, leaving audiences to draw their own conclusions. How do you know a character is female or male? Do you know for certain that they are not non-binary or gender-fluid? Do you know they are not intersex? Do you know they are not trans?

Alternatively, start an analysis from an assumption that all characters in a story are queer, to identify the ways in which scientist characters are coded as heterosexual, cis, or binary by the text. What does it take to prove a character is not queer? Do we know for certain a character is not asexual or aromantic if they have never been depicted in a relationship? Do we know they are not bisexual, pansexual, or otherwise non-heterosexually identifying even if they are in a cis-woman/cis-man relationship? Applying this approach to *Star Trek: Discovery*, we find many more canonically queer or queer ambiguous characters than canonically heterosexual characters. *Discovery* may be exceptional, but the approach can produce surprising results for viewers used to reading fiction with a baseline cis-binary-hetero assumption.

Even among pro-queer fiction texts, pinning down a character's identity can be a problem. Serafini (2017) discusses Cosima Niehaus's scientist girlfriend Delphine in *Orphan Black*, who initially implies she is bisexual, saying 'as a scientist, I know that sexuality is a spectrum.' Serafini argues Delphine's bisexuality is effectively erased when other characters subsequently refer to her as 'lesbian' or 'gay'. The general cultural phenomenon of bisexual erasure might be responsible for this, or it might be sloppy writing, or something else. In our reading, the only time we see Delphine in a relationship with a man she is very uncomfortable, suggesting 'bisexual' may not be an appropriate label. But this insight only emphasizes the point that analysts, like audiences, must decide how to respond to this ambiguity since we cannot ask Delphine herself. Being conscious of the basis for our decisions is crucial to move analyses of scientist characters beyond simplistic and prejudicial assumptions.

Queer audience interpretations

This leads to another issue for science communication analysts and fiction creators alike to consider: how audiences interpret scientist characters.

Queer audiences are used to engaging with fiction from a marginalized position, so their readings will often differ from those of cis-binary heterosexuals (Benshoff, 1997). For example, they may side with the villain of a story, not the hero, if the story perpetuates unfavourable ideologies. They will frequently read characters as queer when this would never occur to cis/straight audiences. Sometimes this is a matter of desire and preference, as with fanfiction writers who find queer angles to a relationship that is not canonically queer. Or it is a matter of different life experiences influencing which aspects of a character each of us reads as signalling queerness. Thus, some read a character as trans, non-binary, intersex, bisexual, or asexual, when other audience members do not. Often such readings are a necessity, because there were so few canonically queer characters until recently, especially scientist characters. All of us want to see ourselves in the fiction we engage with and at times that relies on novel interpretations.

However, it should not just be read as wishful thinking. When audiences identify queer coding in characterizations, they are not reading things into a text that are not there. As Benshoff (1997, p 15) states, such readings 'result from the recognition and articulation of the complex range of queerness that has been in popular culture texts and their audiences all along.' Thus, queer audiences are picking up on real phenomena of at least three kinds.

The first is the real-world existence of diverse kinds of queer people, including people whose identities change, people who are queer 'eggs' (that is, not yet hatched), people who are closeted, and people who keep aspects of themselves private in certain circumstances. To posit that a fictional character is queer but transitioning, closeted, or private is not a big stretch of the imagination; it echoes real-world queer experiences. This is especially the case for STEM professionals who may keep their queerness to themselves in a hostile working environment (see Chapters 6 and 7).

The second phenomenon is the media trope of queer-coding characters to signal there is something wrong with their ideological or moral perspective. This goes beyond scientist characters to all manner of villains (Austin, 2020), but in science-themed fiction it is often relevant to the text's intended messages about science. As Benshoff (1997), Orthia (2011), and Orthia and Morgain (2016) identified for evil and non-credible scientist characters, queer-coding has been used to criticize undesirable kinds of scientific research and to police which kinds of labour may bear science's name. Queer audiences recognize this important function of queer-coding where others may not, potentially providing unique insights into a work of fiction's ideological persuasion.

The third phenomenon is factors that constrain creators of fiction from labelling characters as queer, especially creators of big budget television and film subject to studio veto. Queer audiences are aware of pervasive cultural queerphobia and engaging in queer readings is a way of offsetting this mandated exclusion (Serafini, 2017). Unfortunately, this can result in hostility among fans debating a character's gender, sexuality, or sex characteristics. So long as fiction's creators hide behind a shield of deniability – avoiding queerphobic backlash by refusing to confirm or deny that a character is queer – they are letting queer audiences do the heavy lifting on queer visibility. Audiences are left in interpretative limbo where they know what they can see but their insight is not validated. Worse, creators may get away with producing queerphobic representations themselves – which may be obvious to queer people but denied by other audience members – if they refuse to comment on a character's queer status or suggest a character is not queer while aspects of representation imply they are. Science communicators can play an active role in changing these dynamics by taking queer interpretations of fiction seriously.

Conclusions: implications for science communicators

Science communication as a discipline and profession is nothing without a well-developed understanding of audience, so it is important to incorporate audience interpretations into our understandings of fiction. Fiction does not shape audience views in predictable ways; its impact varies with context and the same text can even affect different people in opposite ways (Orthia, 2019). Clearly we must ask people how they interpret fiction to assess its influence.

For creators of science-themed fiction wanting to support queer audiences, this may mean validating queer audience readings of one's work even if they are different to what was intended. Queer media studies must also incorporate queer audience perspectives. Few published science communication studies have consciously sought queer people's views on fiction, Smith and Tucker's 2020 study being perhaps the only one. More are needed. This is particularly important because of Western science's long history of anti-queer ideologies (Roberson and Orthia, 2021). Science communicators have much to learn about the complex interrelationship of anti-science and pro-science sentiment among queer communities, and studies of queer characters in science-themed fiction can be a fruitful way into that discourse.

While fiction can be complex and people will respond differently to it, queer representation in science-themed fiction remains important at a very simple level. There is a world of difference between fiction that is ambiguously open to interpretation and fiction that explicitly recognizes, names, and validates queer identities. That validation may make a difference to binary, cis, and straight STEM professionals' attitudes to queer colleagues. It will certainly make a difference to queer STEM professionals' sense of wellbeing and potentially to their sense that things could get better. Fiction is a discursive resource that many of us may draw on to help understand and explain the world and our place within it (Kitzinger, 2010). As such, it can be consciously harnessed as a tool for social change. And that aim, after all, is what makes science communication worthwhile.

References

Austin, S. (2020) 'Children of queer bodies: Disney Channel original movies as social justice narratives in *Descendants 2*', in C.E. Bell (ed) *Disney Channel Tween Programming: Essays on Shows from Lizzie McGuire to Andi Mack*, Jefferson: McFarland, pp 255–270.

Benshoff, H.M. (1997) *Monsters in the Closet: Homosexuality and the Horror Film*, Manchester: Manchester University Press.

Bradford, K.T. (2014) 'Invisible bisexuality in Torchwood', *Apex Magazine* [online] 4 March, Available from: https://apex-magazine.com/invisible-bisexuality-in-torchwood/ [Accessed 8 August 2021].

Flicker, E. (2003) 'Between brains and breasts – women scientists in fiction film: on the marginalization and sexualization of scientific competence', *Public Understanding of Science*, 12: 307–318.

Haynes, R.D. (1994) *From Faust to Strangelove: Representations of the Scientist in Western Literature*, Baltimore: Johns Hopkins University Press.

Intersex Australia (2019) 'Intersex population figures', *Intersex Human Rights Australia* [website] 16 September, Available from: https://ihra.org.au/16601/intersex-numbers/ [Accessed 27 October 2021].

Kerry, S.C. (2019) 'Australian queer science fiction fans', *Journal of Homosexuality*, 66(1): 100–116.

Kitzinger, J. (2010) 'Questioning the sci-fi alibi: a critique of how "science fiction fears" are used to explain away public concerns about risk', *Journal of Risk Research*, 13(1): 73–86.

Lavender III, I. (2011) *Race in American Science Fiction*, Bloomington: Indiana University Press.

Long, M., Steinke, J., Applegate, B., Lapinski, M.K., Johnson, M.F., and Ghosh, S. (2010) 'Portrayals of male and female scientists in television programs popular among middle school-age children', *Science Communication*, 32(3): 356–382.

Orthia, L.A. (2010) 'Queer scientists in television science fiction', *Diffusion* [blog] 27 January, Available from: https://lindyorthia.com/2013/12/03/queer-scientists-in-television-science-fiction/ [Accessed 8 August 2021].

Orthia, L.A. (2011) 'Antirationalist critique or fifth column of scientism? Challenges from *Doctor Who* to the mad scientist trope', *Public Understanding of Science*, 20(4): 525–542.

Orthia, L.A. (2019) 'How does science fiction television shape fans' relationships to science? Results from a survey of 575 *Doctor Who* viewers', *Journal of Science Communication*, 18(04): A08.

Orthia, L.A. and Morgain, R. (2016) 'The gendered culture of scientific competence: a study of scientist characters in *Doctor Who* 1963–2013', *Sex Roles*, 75, 79–94.

Penley, C. (1997) *NASA/TREK: Popular Science and Sex in America*, London: Verso.

Roberson, T. and Orthia, L.A. (2021) 'Queer world-making: a need for integrated intersectionality in science communication', *Journal of Science Communication*, 20(01): C05.

Serafini, V. (2017) 'Bisexual erasure in queer sci-fi "utopias"', *Transformative Works and Cultures*, article 24.

Smith, J. and Tucker, B. (2020) 'An in-depth study of *Orphan Black* and its influence upon audience members in relation to their perception of science', Paper presented at the *Australian Science Communicators Eleventh National Conference*, Melbourne: 16–19 February 2020.

StarTrek.com (2020) '*Star Trek: Discovery* introduces first transgender and non-binary characters', *Star Trek* [website] 2 September, Available from: https://intl.startrek.com/news/star-trek-discovery-introduces-first-trangender-and-non-binary-characters [Accessed 8 August 2021].

Steinke, J., Applegate, B., Lapinksi, M., Ryan, L., and Long, M. (2012) 'Gender differences in adolescents' wishful identification with scientist characters on television', *Science Communication*, 34(2): 163–199.

Vary, A.B. (2020) 'Inside the groundbreaking "*Star Trek: Discovery*" episode with trans and non-binary characters', *Variety* [online], Available from: https://variety.com/2020/tv/news/star-trek-discovery-trans-non-binary-blu-del-barrio-ian-alexander-1234824183/ [Accessed 26 July 2021].

Practice Spotlight: Using #QueerInSTEM and Related Hashtags to Promote Your Science Communication

Luis Lopez and Alberto I. Roca

Social media is a common forum for engaging fellow scholars as well as different publics on science topics. What strategies do queer scientists and communicators use with social media to network with one another and to reach a public audience? We address this topic through our analysis of queer-related STEM posts on Twitter. This reviews activities by practitioners that can serve as lessons about and for queer science communication (SciComm). For this project, we monitored ten LGBTQ and STEM connected hashtags and collected over 4,500 tweets published by more than 2,000 unique user accounts during a 12-month period from June 2020.

Overall, among the hashtags that we had searched, the most frequently used were the following (in decreasing order): #QueerInSTEM (>30 per cent of tweets contained this tag; see Figure Sp7.1), #LGBTQinSTEM, #LGBTQSTEM, and #LGBTinSTEM. Some Twitter users included other tags, that we did not intentionally monitor, to represent either specific populations within the queer community (#TransInSTEM, #BiInSci) or special events (#PrideInSTEM, #LGBTQSTEMday). The latter two hashtags coincide with peak utilization of these labels. Namely, the Pride celebratory month of June in the USA as well as the International Day of LGBTQIA+ People in STEM (18 November) may be when there is a potentially larger audience for your posts. There were also additional tags that represented intersectional identities of the tweet user or of their intended audience such as #WomenInSTEM, #BlackInSTEM, #LatinxInSTEM, #DisabledInSTEM, and #NativeInSTEM. Therefore, we recommend labelling your posts with these additional tags to facilitate content discovery online and to connect with your desired intersectional queer STEM community.

Besides hashtags, another way of finding a queer SciComm topic is by following the accounts that most often tweeted with the aforementioned tags. In our dataset, the top institutional accounts representing queer groups were @TheSTEMvillage (97 Tweets), @LGBTSTEM, @PRISMexeter, @Sci_QUEERies, @STEMEquals, @QueerSciencePH, and @OUTinSTEM.

The content of these hashtag-containing tweets may be more informative as examples for how to queer SciComm. The most frequently used (non-hashtag) words were related to the sciences and diversity: STEM (used 656 times), WomenInSTEM, Pride, LGBTQ+, and science. More importantly, a preliminary sentiment analysis of the entire dataset determined that the words 'happy', 'support', and 'love' were the top positive words. By contrast, 'issues', 'discrimination', and 'harassment' were in the top ten negative words. The collection of tweets was in general more positive in sentiment than negative. Such negative sentiment may reflect Twitter users sharing stories of harassment and discrimination.

We tentatively identified themes within the dataset. Some tweets[1] engaged in resistance and advocacy as demonstrated by @EndoKweer:

> Just learned that JKR uses Robert Galbraith as a pen name. Everyone in STEM should know this heinous person. Dr Robert Galbraith Heath tried to "cure" homosexuality with deep brain stimulation. This cannot go unaddressed #queerinSTEM #AcademicTwitter #AcademicChatter

This user is the University of Michigan postdoctoral researcher Dr Daniel Pfau who writes a *Psychology Today* blog called 'Queering Science' where they had published a related post (19 June 2020).

Another theme shows how social media allows users to access the 'hidden curriculum' in academic professional development such as when @TheDoctaZ asks:

> Hey #queerinstem @500QueerSci Soooo ... first publication. Do I do legal name or preferred name? Is it more valuable to be searchable or avoid being misgendered for all eternity? #phdchat #sacnas2020 #ostem2020

Finally, sharing on social media through 'shout outs', 'roll calls', and other posts representing your identity and experience gives testimony to the existence of the academic queer STEM community. Sometimes this can be achieved through witty and geeky science humour as exemplified by @NathanLents:

Figure Sp7.1: Word cloud showing the frequency of hashtags in the LGBTQ and STEM dataset

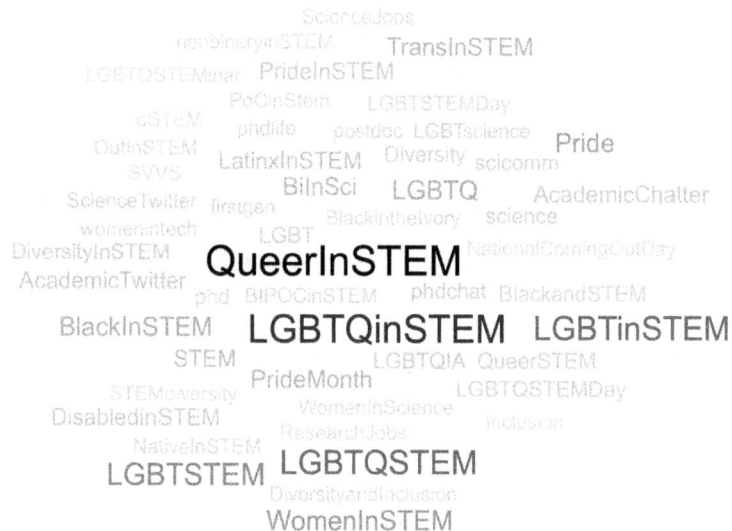

Even as a biologist, #FatBearWeek was just not what I was expecting it to be. #LGBTQinSTEM

Such testimonials can also document the fellowship that an online support network can provide as shown by Em Haydon (@emhaydon):

> Look at all those lovely friendly faces! Getting back into twitter (esp #scicomm and #QueerinSTEM twitter) has been one of the best things I've done this year. Helped me feel part of a community whilst on furlough.

Haydon's tweet included a collage of profile photos from their network illustrating the connections that queer SciComm can achieve.

Note
[1] All tweets quoted with permission.

Practice Spotlight: Queer Science Blogs: Public Communication Before the Age of Social Media

Ron Buckmire and Alberto I. Roca

Before the ubiquity of social media, people interested in sharing their thoughts and experiences on the internet mostly relied on blogging. Early STEM queer-related blogs of this era included *Denim and Tweed* authored by Jeremy Yoder. Under the 'Diversity in Science' blog carnival series (Roca and Yoder, 2011), Yoder also hosted the first carnival celebrating LGBTQ science and scientists, which serves as a sample of blog posts from this time (*circa* 2000–2010) (Yoder, 2011). This spotlight reflects on a blog created by the first author devoted to STEM and queer issues, *The Mad Professah Lectures* (Buckmire, 2022).

In the beginning around 2003–2005, Buckmire blogged about queer politics and forthrightly declared his positionality in the blog description as 'a Black, Gay, Caribbean, Liberal, Progressive, Moderate, Fit, Geeky, Married, College-Educated, NPR-Listening, Tennis-Playing, Feminist, Atheist, Math Professor in Los Angeles, California.' His earliest blog posts were contemporaneous musings about the United States Supreme Court's landmark ruling *Lawrence v. Texas* in which the high court ruled all existing anti-sodomy laws unconstitutional, proclaiming that it was granting gay men and lesbians equal rights under the law for the first time.

Buckmire's blogging was an early online outlet for queer science communication with posts tagged such as 'science', 'STEM', and 'mathematics'. More broadly, *The Mad Professah Lectures* was intended for a general audience since his blog posts included pictures of shirtless, scantily-clad, muscular men (tagged 'Eye Candy'), atheism (Godless Wednesday), celebrities (Celebrity Friday), LGBTQ people (Queer Quote), book and movie reviews (Reviews), and obsessive cataloguing of men's and women's grand slam tennis matches (Tennis).

However, once at the blog, readers would also learn about important science policy issues such as HIV/AIDS information, STEM public policy (especially related to mathematics), and diverse representation and participation in STEM. One of the most impactful science communication topics covered was the issue of the criminalization of HIV in the context of the infamous 'Tiger Mandingo' case. Tiger Mandingo was the eye-catching sobriquet allegedly used by Michael Johnson, a former college wrestler, who at age 21 in 2013 was arrested and charged by the state of Missouri with 'recklessly infecting' and 'recklessly exposing' multiple male sexual partners to HIV while being HIV positive himself. Buckmire first blogged about the case within days of the initial press reports in 2015, when Johnson was found guilty at trial. Buckmire wrote about the anguish and surprise at the draconian punishment (from 30 years to life in prison) that a fellow Black gay man could receive for failing to disclose his HIV status. The focus on Michael Johnson's predicament shows how Buckmire used the medium of blogging as an emotional and creative outlet that also reached readers seeking community and commonality.

Buckmire's blog is just one example among many of the ways that web-based communication was affected by and had impacts on 'offline' settings. The Michael Johnson case eventually became a *cause célèbre* in Black gay circles and in the wider LGBTQ media and blogosphere because of several factors. The story had multiple interesting angles: sex appeal, injustice, discriminatory public policy, and courtroom drama. Activists, who had long advocated an update of 1980s era legislation on 'exposure to AIDS', saw the case as a media friendly vehicle to educate the public, policymakers, and the press about the stakes involved in HIV criminalization. In 2015, when Johnson was sentenced, Buckmire continued to cover the story (Buckmire, 2015). He directed readers to the Center for HIV Law and Public Policy, an organization working against HIV criminalization. This story had an influence on Buckmire's real-life work as an LGBTQ activist. For several years he was on the board of directors of the Center for Health Justice, an organization that advocates for people incarcerated in Los Angeles County to receive access to education, support, and treatment related to HIV/AIDS and other serious conditions.

In summary, Buckmire's blog was an outlet for a queer STEM professional who was involved in various activist movements in the early 21st century. Buckmire's blogging exemplified the adage that 'the personal is political'. This blog was a pioneering example of queer science communication that helped pave the way for the thousands of out, queer STEM professionals that are active on social media today. Being out as queer in STEM professions is still not easy, especially for Black people and other people of colour. Thus, online visibility is a crucial way to undermine stereotypes and to remind diverse people that we all have a place in science while being our authentic selves.

References

Buckmire, R. (2015) 'UPDATE: Black Gay Man, 23, Sentenced To 30 YEARS Under Missouri's HIV Criminalization Law', *The Mad Professah Lectures* [blog] 13 July, Available from: https://buckmire.blogspot.com/2015/07/update-black-gay-man-23-sentenced-to-30.html [Accessed 20 March 2022].

Buckmire, R. (2022) 'The Mad Professah Lectures', *The Mad Professah Lectures* [blog] 31 March, Available from: https://buckmire.blogspot.com [Accessed 20 March 2022].

Roca, A.I. and Yoder, J.B. (2011) 'Online LGBT Pride: diversity in science blog carnival', *DiverseScholar* 2: 1.

Yoder, J.B. (2011) 'Diversity in Science Carnival: Pride Month 2011', *Denim and Tweed* [blog] 30 June, Available from: https://denimandtweed.jbyoder.org/2011/06/diversity-in-science-carnival-pride-month-2011/ [Accessed 20 March 2022].

Teaching Notes for Part II

Visibility matters for LGBTIQA+ folk. Overt inclusion of rainbows and other signs of inclusion mean a space might be welcoming and safe. Other ways of demonstrating support include joining and promoting your institution's LGBTIQA+ ally programme (if you have one) and using your pronouns to introduce yourself on your teaching materials. How can you signal your support of LGBTIQA+ students in your classroom?

It is also good to reflect on how structural elements of teaching might be difficult for your students. For instance, make sure any sign-in sheets for your class give students the option to use their preferred name if it is different from their legal name. This is especially important for trans, non-binary, and gender diverse students if they carry a legal name they do not identify with, often called a 'dead name'. Be careful not to use students' dead names. To do so can misgender them, out them, offend them, and alienate them from you and your class.

Questions posed by Part II:

- **Interrogating assumptions:** what are the default assumptions communicators have built into science communication content such as a fiction film or science museum exhibit? Is there a presumption of heterosexuality or cisnormativity embedded within it? How can we disrupt those assumptions to include more diverse perspectives? How might that make our science communication practice better?
- **Channels for queer perspectives:** where is queerness visible within the science communication world? To what extent is it marginal, and to what extent mainstream? Why might queer science communicators use social media to communicate science with a queer inflection? Where are queer people's views about science and science-based content most likely to be heard, and why?
- **Audience selection:** we know that science communication in the past has been limited in terms of who it reaches and when. What steps can you take to diversify your audiences and communication partners, in particular reaching more LGBTIQA+ people? How can you support

contributions from participants of diverse identities, professions, and linguistic backgrounds?

Activities:

- **Text analysis:** critically analyse the assumptions that underlie dominant scientific theories relevant to sexuality, gender, and sex characteristics, such as Charles Darwin's sexual selection theory. What assumptions inform those theories? How do they limit our understanding of the non-human world? What are the implications of that for understanding human experiences? How should science communicators explore and address those limitations?
- **Field trip:** visit a public STEM institution, for example, a museum or a zoo. Ask your students to pick a display to review in relationship to queering science communication. What does the display communicate? And what does it leave out? Design or suggest improvements for the display.
- **Science communication in drag:** play a video in class of a drag-based science communication performance. Discuss your responses to it as a group. How does the drag approach enhance the act of communicating science? Did the performance challenge you, in terms of your thoughts and feelings about gender, sexuality, or science? In an extended activity, invite (and pay) your local drag performer to run a workshop in your classroom on the basics of drag performance. Work with students to produce your own drag performance that communicates science entertainingly.

PART III

Queer People in Science Communication Communities

6

Malayang Paglaladlad para sa Mapagpalayang Paglalahad: Coming Out and Queering Science Communication in Contested Spaces

John Noel Viaña, Mario Carlo Severo, Miguel Barretto-Garcia, Paul James Magtaan, Jason Tan Liwag, Roemel Jeusep Bueno, Christer de Silva, and Ruby Shaira Panela

Introduction

Many LGBTIQA+ individuals struggle with pervasive discrimination, prejudice, and stigmatization, detrimentally affecting their physical and mental health (Clark, 2014). Although science communication can present these concerns, and even encourage action, it may be limited by its scientific focus, which sidelines queer voices and objectifies queer individuals (Roberson and Orthia, 2021). This exemplifies the urgent need for queering science communication theory and practice, challenging traditional cis-heteronormative visions, and including diverse voices for genuine social transformation (Rumens et al, 2018). Furthermore, diverse modalities of queer science communication are needed, especially in contested spaces where queer bodies and identities are constantly being questioned, discriminated against, and marginalized; and where Western views of queering science communication may not fully acknowledge local culture, history, and socioeconomic condition. In this chapter, we use the Philippines as a lens to explore queering science communication in a non-Western space where the queer and the colonized remain subjugated and dispossessed by social, economic, and cultural conditions.

We first illustrate how the Philippines is a contesting space for queer identities and then reflect on opportunities and barriers for queering science communication in the country. In order to explore challenges in coming out as a queer Filipino scientist/science communicator and their implications for communicating science that impacts queer people, we employ collective autoethnography. We draw upon and jointly reflect on our lived experiences (Chang et al, 2013), illustrating these challenges as 'subjective eyewitnesses' and 'experienced knowers' with discourses of authenticity and suffering (Jasanoff, 2017). We conclude by laying the groundwork for what queering science communication entails and how it can promote free expression and societal equity in vulnerating spaces for the LGBTIQA+ community.

The Philippines: a contested space for queer identities

The Philippines is an archipelagic Southeast Asian nation composed of diverse indigenous and ethnolinguistic groups. It has endured Spanish colonial rule for 300 years and the US for half a century (Tan, 2001). Prior to colonial subjugation, effeminacy, cross-dressing, and gender-transitive behaviours were observed in the indigenous culture (Garcia, 2013; UNDP and USAID, 2014), epitomized by the *babaylans* or *bayogs*. They are ancient male spiritual leaders (Garcia, 2013) who assume a spiritual intermediary role by performing rituals while adopting a feminine demeanour and wearing customary female clothing (UNDP and USAID, 2014). *Babaylans* enjoyed a symbolic status as *binabae* (or 'womanlike') and engaged in feminine work, with others even having husbands and conjugal relations with men (Garcia, 2013).

Spanish colonization brought the disappearance of *babaylans* and the introduction of Catholicism (UNDP and USAID, 2014), shaping prevalent attitudes and social expectations on identity, gender, and sexuality. Catholic dogmas implanted *machismo* and conservative family values in the Filipino culture (Nadal and Corpus, 2013). Individuals who did not adhere to conventional gender identity and heteronormative relationships were shunned or shamed.

At present, deep-seated religious norms from Spanish colonization pervade, with same-sex marriage and sex reassignment still not legally recognized (Manalastas and Torre, 2016). Despite being a signatory to international human rights conventions (UNDP and USAID, 2014), the Philippines is yet to ratify its Sexual Orientation and Gender Identity Expression (SOGIE) Equality Bill for combatting LGBTIQA+ discrimination. Two decades after its proposal, the enactment of the SOGIE Equality Bill remains, unfortunately, a question.

Intersecting contested contexts: impact on queering science communication

Currently, queer individuals in the Philippines face multiple challenges as a result of the country's colonial history and socio-political situation. For queer Philippine science communication to promote queer rights, expression, and equity, it is crucial to first unpack how various LGBTIQA+-contesting features of the Philippine context intersect and impact both queer scientists and queer science communicators.

First, queer science communication needs to acknowledge concerns of invisibility of the LGBTIQA+ community in science (Roberson and Orthia, 2021). Currently, there is limited space for queer Filipino scientists to exist. Though collective efforts such as Queer Scientists PH (Liwag, 2021) have initiated discussions surrounding the struggles and triumphs of queer Filipinos in STEM, there are still limited publicly available institutional research and structural policies that protect queer science contributors. With sparse queer representation in mainstream Filipino media (Inton, 2017; Yorong et al, 2021), there are minimal opportunities for queer scientists and science communicators to share science relevant to them, particularly ideas challenging dominant cis-heteronormative narratives.

Second, there are limited programmes for dedicated science communication training and minimal government funding available for science communication activities in the Philippines (Navarro and McKinnon, 2020). Although there are increasing numbers of science communication articles discussing agricultural biotechnology, disaster risk reduction, and public health (Montemayor et al, 2020), other areas, including transgender health, remain underrepresented. Moreover, socio-economic disparities limit access to scientific information with science barely present in major broadsheets, television channels, and public radio broadcasts (Navarro and McKinnon, 2020).

Third, marginalized communities are usually not included in scientific endeavours and communication initiatives. When LGBTIQA+ individuals are portrayed, they may simply be delegated as narrative rather than as deliberative agents. Although their stories and struggles are narrated, there are limited opportunities for them to speak and directly deliberate with other stakeholders on issues impacting their health and wellbeing (Curato and Ong, 2015).

Finally, religion still shapes debates regarding public health (Curato and Ong, 2015) in a highly religious country like the Philippines. Whenever pitted with science, religion is considered as always 'right' (Montemayor et al, 2020). The enduring influence of religion on mass communication and public policy presents a barrier for queer science communicators to be open about themselves and their preferences, for fear of being considered as sinful and outcaste.

Paglaladlad and *paglalahad*: personal narratives of 'coming out' and 'telling about'

Situating the lived experiences of LGBTIQA+ individuals in the Philippines's socio-historical context makes *paglaladlad* – the indigenous term for 'coming out' that literally translates to 'unfurling one's cape' (Tan, 2001) – exceptionally challenging. To effectively understand the impact of challenges of *paglaladlad* and *paglalahad* (communicating science) on queer scientists or science communicators, we adopted a collaborative autoethnography approach, collectively gathering and interpreting autobiographical data as sociocultural phenomena (Chang et al, 2013). We draw from the perspectives of scientists who are both producers and transmitters of scientific knowledge, media practitioners who communicate scientific information to various publics, and school teachers who play a primary role in introducing science to the public.

Roe: disclosing, masking, and owning queer identity in the science lab

Queer scientists experience discrimination in the workplace – in laboratories, conference rooms, and offices. Discrimination happens as the researchers navigate their work and while they communicate and interact with their colleagues. Moreover, queer STEM professionals have to negotiate their visibility in research institutions, which remain cis-heteronormative (Yoder and Mattheis, 2016). These phenomena are evident in Roe's experience, where his supervisor was unwelcoming of queerness. Roe describes his experience of coming out at work:

> In 2011, my boss asked if I have a girlfriend. I shyly said, "No Ma'am, I have a boyfriend." I saw her face change from an excited look to a concerned one. She's a devout Christian who ends her e-mails with "God Bless!" She then remarked, "Oh I would never have guessed. I'm sure that's just a phase that you are going through." I froze. I didn't know what to say. And as a fresh university graduate in his first job in the country's centre of excellence in the health sciences, I just nervously smiled, thanked her for her time, and left her office.

Roe's *paglaladlad* was met with a prejudiced remark that made him uncomfortable. His supervisor implied disapproval of his sexual orientation by alluding that 'it is just a phase', delegitimizing an integral part of Roe's identity and dismissing his capacity to define his sexuality (Toft et al, 2020). By initially asking if he has a girlfriend, his supervisor set heteronormative expectations for members of the lab. Conservatism contributes to the fear of coming out for gay people (Kurzweil, 2008), and when compounded by power imbalances (Toft et al, 2020) between supervisor and student, can make not only 'coming out', but also 'existing as' challenging. The person disapproving of Roe is his supervisor – someone with profound influence on his career progression in the highly hierarchical Philippine society. Roe's vignette also demonstrates how the Philippines' highly religious environment can make queer people fearful of coming out, delegitimize their identities, and question their capability for self-determination.

Roe's vignette also touches on the intersectionality of his identity and how it affects his research and communication with colleagues. He is a researcher, and he is queer. However, after the incident with his supervisor, he became more cautious in interacting with colleagues, compartmentalizing what he deemed as 'professional' from 'personal'. Many times, he would be worried about his safety and his colleagues' perception of him, pushing him to retreat and project himself in a heteronormative way:

> My outlook on other employees got impaired. The fact that my boss reacted that way made me fearful that other employees in our research institution would also react similarly, or even worse. I always had to kind of tiptoe around certain employees that I assume have a conservative background. I also caught myself behaving in a heteronormative way so that other employees would not suspect that I am queer. I thought of just maintaining a 'professional' image at work and only being loud and proud outside of work.

Roe's experience reflects how presenting himself in a heteronormative way served as a straightjacket in his scientific communication with peers. He added:

> As much as I do not want my professional scientific work to be affected by the incident, I became very mindful of how I present my research. When entertaining laboratory visitors and presenting in scientific conferences, I tend to mask my queerness by speaking in a lower voice tone and avoiding unnecessary hand gestures. I felt like if the other academics will know that I am gay, they wouldn't take my research seriously or worse, would discredit it.

Roe tailors his interactions with colleagues on how they may perceive him, in fear of being judged as queer and its effects on how his work is perceived. The compartmentalization and communication accommodation by Roe towards his colleagues (Dragojevic et al, 2015) took a toll. His actions imply that he is invisibly queer in the workplace, which could have negative repercussions for his mental health, productivity, and career advancement (Powell et al, 2020).

> I always feel exhausted mentally and physically for just putting up a 'straight' front and always being very aware of how I should act in a professional setting. I started to not be excited when having a meeting with certain people in my institution, and also just having other academics as guests in our lab. I love doing science, but not being my authentic self was very tiring and painful.

Eventually, when Roe felt safe and more comfortable with his colleagues, he began to deliver better outputs and became more open in interacting. Roe said:

> After I started changing my mindset, I noticed also a change in the people around me. Other employees got to know me more easily since there were events in the institute that I was asked to host. On these occasions, I always take the opportunity to let other employees know that I am queer.

Jason: fostering agency in online queer science storytelling

The internet has become a gateway for understanding one's queerness – a third space for exploring sexual orientation, gender identity, and gender expression. Online anonymity provides many LGBTIQA+ individuals an opportunity to authentically express themselves beyond their homophobic and transphobic localities (Dhoest and Szulc, 2016). Unfortunately, the internet has also become a platform for Jason's premature 'outing':

> Being out online does not equate to being out offline. With the veil of anonymity removed, queer lives are easily threatened by real-world consequences of their queerness. In an attempt to honour my work as a scientist, I was outed by one of the largest gay magazines in the world. Failing to ask me for permission prior to publication, they unexpectedly and involuntarily accelerated the process of coming out to my family.

Although showcasing queer scientists and their scientific achievements could be well-intentioned, the international online magazine demonstrated lack of regard for Jason's situatedness in a country with limited rights and acceptance for queer individuals. This highlights the challenge in showcasing queer identities in science communication – efforts towards celebrating diversity must largely consider one's context and respect one's agency.

Whereas diversity is increasingly being celebrated in Western culture, queer visibility still has to be complexly negotiated with familial and social ties in other more conservative and less accepting cultures. In the latter contexts, outing can deprive individuals of their dignity, privacy, and autonomy, and could hurt, rather than help, individuals and queer communities (Guzman, 1995). Without asking for Jason's consent, the online magazine infringed his privacy and agency. His being out in certain online spaces does not necessarily equate to him also being out elsewhere, and deliberately outing him in platforms visible to the general public compromises his choice to only be open in spaces where he feels safe and comfortable. At some point, Jason even questioned whether his gay magazine feature deserves a celebration, as he could not share it to his family. He reflected on the incident's aftermath, personally and professionally:

> There are social relationships necessary to maintain for science communication to even occur and these are unfortunately, at times, threatened by disclosure of one's SOGIE. I was lucky enough to be accepted by my family, and it gave me the confidence to publicize my experience. People began reaching out to me not only for science communication engagements but also in solidarity and support. But there have been others before me who have been outed by magazines and thrown out of their homes, fired from their jobs, and/or worse.

Jason's experience demonstrates the need for queer people to be able to exercise both narrative and deliberative agencies (Curato and Ong, 2015), deciding when, where, and how to share their stories. They should be given the chance to appraise the costs and benefits of coming out (Guzman, 1995), and be free to decide to whom will they be out. As such, improving queer visibility in science communication requires 'strategic outness' – a constant and contextual management of one's queer identity (Orne, 2011). For effective *paglalahad* to occur, *paglaladlad* must be conducted in a manner that is sensitive to the individual's social context and does not compromise one's autonomy, privacy, and safety.

From his experience, Jason vowed to ask for consent when featuring queer scientists on his Philippine-based online advocacy platform, Queer Scientists PH (see Practice Spotlight: Queer Scientists PH). He informs contributors when they will be featured on its Instagram and Twitter accounts and also provides them the opportunity to review and revise their interview statements before they are posted online. The platform also allows scientists to deliberate on pertinent matters by encouraging them to actively engage in any discussion. Additionally, contributors can always update their profiles and even withdraw their contributions, retaining ownership and power over their narratives.

Jason's experiences taught him how social media can promote the social relevance of his scientific work and enable queer lives to be shared to the public. Despite being outed on social media, he used the same platform to take ownership of his narrative, through his article on CNN (Liwag, 2021). He further reflects,

It is a privilege to be able to write about my experience on a platform as wide as CNN Philippines, and it made the process of reclamation much easier by allowing me to address the audience directly and hold the gay magazine (which I could not contact) accountable for their oversight. Visibility in itself is not a goal, but rather a means to an end. However, in the absence of these opportunities for narrative reclamation, I do not know if I would have had a chance to continue with science communication afterwards.

Paige: promoting trans visibility in the science classroom

Heteronormativity and cis binary gender expectations are prevalent in schools (Endo et al, 2010), especially in the Philippines where there is a deeply entrenched cis-heteronormative culture. Challenging institutionalized cis-heteronormativity in pedagogical spaces requires queer visibility and queering science education/communication. Queer educators should be encouraged to come out and teach as their authentic selves. Unfortunately, coming out is an uphill battle for many queer educators and can put their safety and job security at stake.

Paige's story captures the struggles of queer educators in coming out in schools, especially when there is limited enforcement of anti-discrimination policies:

> As the lone transgender teacher in a public school, my coming out as a trans woman was met with ridicule and repulsion. When I started to subtly change my appearance, my co-teachers would ask 'Why do you grow your hair long – isn't there a prescribed haircut for male teachers?' Even my 'straight-acting' gay colleagues asked, 'Why can't you just be a gay man? If the office finds out, they might reprimand you.' I felt alone, filled with fear and anxiety. I was worried that I wouldn't be allowed to come out, or that my colleagues would report me and I would then be removed from teaching. I then decided to isolate myself, hoping that I would be invisible. I would hurry home after class to evade comments like 'What will the student's parents say knowing that their adviser is a trans woman? The students are young and in their primary level. The children might imitate their teacher.' I would even skip school events, even if I wanted to participate.

Paige's transition was met with hostility. Micro-aggressions from colleagues prompted Paige to isolate and denied her of career advancement opportunities. Nevertheless, Paige vowed to live her reality:

Despite these fears and anxieties, I continued with my transition and eventually went for sex reassignment surgery. I want to be seen, known, treated, and accepted as a woman – because I'm a woman.

For Paige, coming out was never a single process, but a repetitive one. As a teacher, each new school year means new students with the same questions on past identities. As with many transgender individuals, her coming out not only demonstrates the process as ongoing and socially embedded (Brumbaugh-Johnson and Hull, 2018), but it also stresses the importance of educating people about queer lives and experiences:

On the first day of each school year, I would write my assigned name at birth on the board and would explain that I prefer to be called 'Teacher Paige or Ma'am Paige'.

This yearly introduction started when a group of boys mocked me, which made me feel furious and disrespected. While the students eventually apologized, other tête-à-têtes became a handful, so I decided to address it openly in the classroom. It wasn't easy facing each new cohort and explaining the same thing over again. It is exhausting. But by sharing my preferred name and my 'true' identity, I hope that my students and colleagues would treat me well.

Paige's sharing of her identity to her students has become a yearly ritual. Although burdensome, ensuring that she is addressed as a 'Ma'am/Miss' makes her feel seen, accepted, and respected. Paige's story resonates with the coming out of many transgender individuals, where they have to balance between living their authentic self and dealing with its social demands (Levitt and Ippolito, 2014). Having to constantly explain one's identity causes a psychological toll, and could adversely affect pedagogy.

The lack of gender inclusivity and queer representation in the curriculum complicates Paige's teaching:

While I was presenting a slide on the male and female reproductive systems, I heard my students' chitchat so I paused and told everyone, 'Class, did you know that I used to have a penis? But because of science, I gained a vagina.' My students were astonished. One even asked out loud, probably out of sheer curiosity, 'So, you're REALLY a woman now, Ma'am?' Since biology textbooks don't discuss queer identities, I took the opportunity to introduce my sexuality.

Aside from satiating my student's curiosity, disclosing my gender frees me of the burden of pretension. It gives me space to build more meaningful relationships in school – knowing that I am truthful within my teaching environment. Opening up also provides me an opportunity to deepen my students' understanding of the struggles and experiences of the LGBTIQA+ community. Transgender people exist, and we can become effective teachers. And when my students tell me that they become mindful in treating queer individuals, I count those moments as success.

Drawing from Paige's experience, support for the *paglaladlad* of queer educators should be an agenda for queering science communication to ensure effective *paglalahad*. There should be institutionalized policies that protect queer educators and students from discrimination. Additionally, gender sensitivity and inclusivity in school curricula and teacher trainings should be strictly mandated and closely monitored for compliance. Implementing these policies would create a safe atmosphere in schools, fostering queer visibility and representation, and ensuring that scientific topics relevant to queer people are taught properly and sensitively.

Paglaladlad para sa malayang paglalahad: towards societal equity

Paglaladlad is essential for free *paglalahad*. Open communication on queer science can better happen if queer scientists and science communicators can tell facts, stories, and struggles as their authentic selves. In turn, this can lay the foundations for societal equity by foregrounding faces behind scientific facts; empowering queer individuals to narrate and deliberate relevant health, scientific, and policy issues (Curato and Ong, 2015); and providing venues to rally for their rights in the very societies that contest the validity of their identities.

Indeed, queering science communication can be challenging in contexts with a rich queerphobic history driven by a colonial legacy of religion and patriarchy. *Babaylans* are no longer revered, and the *bakla*, *bayot*, or *tomboy* (local terms for queer people) are often ridiculed. As much as we would like to illustrate diverse queer experiences, our discourse is limited by our relatively privileged position as people with research or teaching roles. Future analyses should also explore the stories and struggles of other marginalized queer individuals, including those with limited education or technology access and those in liminal spaces, such as diasporic communities (Barretto-García, 2021); and determine how they could benefit from queer science communication.

Overall, this chapter lays the foundations for further discussions on queering science communication in the Philippines and in other spaces where queer identities are continuously being contested and invalidated. With homosexual behaviour still criminalized in 71 countries at the time of writing

(Gerber et al, 2021), queering science communication can help set the stage for challenging existing policies and promoting queer rights and welfare. It can even encourage community building and provide opportunities for queer persons to collaborate, coordinate, and demand change. But for these to happen, queer scientists and science communicators must be given the opportunity to 'unfurl their capes' at their own time and terms. They must not only be seen, but be heard.

Dedication and acknowledgement
In memory of all the people we have lost along the way. In loving memory of Dennis Schmidt, partner of Roemel Bueno, who selflessly shared his light to the world. In memory of Prof. Alonzo A. Gabriel, who paved the way for many Filipino queer scientists to express, live, and be proud of their authentic selves.

We also thank Kamila Navarro for her help in forming the team.

References
Barretto-García, M.B. (2021) 'The poem annotates power', *Dark Horse* 43: 12–20.
Brumbaugh-Johnson, S.M. and Hull, K.E. (2018) 'Coming out as transgender: navigating the social implications of a transgender identity', *Journal of Homosexuality*, 66(8): 1148–1177.
Chang, H., Ngunjiri, F.W., and Hernandez, K-A.C. (2013) *Collaborative Autoethnography*, London: Routledge.
Clark, F. (2014) 'Discrimination against LGBT people triggers health concerns', *The Lancet*, 383(9916): 500–502.
Curato, N. and Ong, J.C. (2015) 'Inclusion as deliberative agency: the selective representation of poor women in debates and documentaries about reproductive health', *Television & New Media*, 16(6): 576–594.
Dhoest, A. and Szulc, L. (2016) 'Navigating online selves: social, cultural, and material contexts of social media use by diasporic gay men', *Social Media + Society*, 2(4): n.p.
Dragojevic, M., Gasiorek, J., and Giles, H. (2015) 'Communication Accommodation Theory', in C.R. Berger, M.E. Roloff, S.R. Wilson, J.P. Dillard, J. Caughlin, and D. Solomon (eds) *The International Encyclopedia of Interpersonal Communication*, New Jersey: Wiley-Blackwell.
Endo, H., Reece-Miller, P.C., and Santavicca, N. (2010) 'Surviving in the trenches: a narrative inquiry into queer teachers' experiences and identity', *Teaching and Teacher Education*, 26(4): 1023–1030.
Garcia, J.N.C. (2013) 'Nativism or universalism: situating LGBT discourse in the Philippines', *Kritika Kultura*, 20: 48–68.

Gerber, P., Raj, S., Wilkinson, C., and Langlois, A. (2021) 'Protecting the rights of LGBTIQ people around the world: beyond marriage equality and the decriminalisation of homosexuality', *Alternative Law Journal*, 46(1): 5–12.

Guzman, K.R. (1995) 'About outing: public discourse, private lives', *Washington University Law Review*, 73(4): 1531–1600.

Inton, M.N. (2017) *The Bakla and the Silver Screen: Queer Cinema in the Philippines*, PhD Thesis, Lingnan University.

Jasanoff, S. (2017) 'Virtual, visible, and actionable: data assemblages and the sightlines of justice', *Big Data & Society*, 4(2): n.p.

Kurzweil, J. (2008) 'Shattering the glass closet', *Science* [online] 5 December, Available from: https://www.sciencemag.org/features/2008/12/shattering-glass-closet [Accessed 26 March 2022].

Levitt, H.M. and Ippolito, M.R. (2014) 'Being transgender: the experience of transgender identity development', *Journal of Homosexuality*, 61(12): 1727–1758.

Liwag, J.T. (2021) 'How a magazine feature forced me to come out to my parents', *CNN Philippines Life* [online] 18 January, Available from: https://cnnphilippines.com/life/culture/2021/1/18/coming-out-story-attitude-magazine.html [Accessed 26 March 2022].

Manalastas, E.J. and Torre, B.A. (2016) 'LGBT psychology in The Philippines: initial inroads', *Psychology of Sexualities Review*, 7(1): 60–72.

Montemayor, G.J.S., Navarro, M.J., and Navarro, K.I.A. (2020) 'Philippines: from science then communication, to science communication', in T. Gascoigne, B. Schiele, J. Leach, M. Riedlinger, B.V. Lewenstein, L. Massarini, and P. Broks (eds) *Communicating Science: A Global Perspective*, Canberra: ANU Press, pp 665–692.

Nadal, K.L. and Corpus, M.J.H. (2013) '"Tomboys" and "baklas": Experiences of lesbian and gay Filipino Americans', *Asian American Journal of Psychology*, 4(3): 166–175.

Navarro, K. and McKinnon, M. (2020) 'Challenges of communicating science: perspectives from The Philippines', *Journal of Science Communication*, 19(1): A03.

Orne, J. (2011) 'You will always have to "out" yourself: reconsidering coming out through strategic outness', *Sexualities*, 14(6): 681–703.

Powell, K., Terry, R., and Chen, S. (2020) 'How LGBT+ scientists would like to be included and welcomed in STEM workplaces', *Nature*, 586(19 October): 813–816.

Roberson, T. and Orthia, L.A. (2021) 'Queer world-making: a need for integrated intersectionality in science communication', *Journal of Science Communication*, 20(1): C05.

Rumens, N., De Souza, E.M., and Brewis, J. (2018) 'Queering queer theory in management and organization studies: notes toward queering heterosexuality', *Organization Studies*, 40(4): 593–612.

Tan, M.L. (2001) 'Survival through pluralism: emerging gay communities in The Philippines', *Journal of Homosexuality*, 40(3–4): 117–142.

Toft, A., Franklin, A., and Langley, E. (2020) '"You're not sure that you are gay yet": the perpetuation of the "phase" in the lives of young disabled LGBT + people', *Sexualities*, 23(4): 516–529.

UNDP and USAID. (2014) *Being LGBT in Asia: The Philippines Country Report*, Bangkok: United Nations Development Program.

Yoder, J.B. and Mattheis, A. (2016) 'Queer in STEM: workplace experiences reported in a national survey of LGBTQA individuals in science, technology, engineering, and mathematics careers', *Journal of Homosexuality*, 63(1): 1–27.

Yorong, D., Bacarra, A.K., and Andres, G. (2021) 'Miss Bulalacao: a kineikonic (re)configuration of queer and social spaces', *Plaridel*, 18(1): 203–229.

Practice Spotlight: Queer Scientists PH: Visibility Towards Community Building and Empowerment

Jason Tan Liwag, Jay S. Fidelino, Rey Audie S. Escosio, Almira B. Ocampo, and Nikki Santos-Ocampo

Queer Filipino STEM workers exist within a space of triple marginalization due to poor scientific research conditions and discrimination due to race and sexual orientation, gender identity, and gender expression (SOGIE). Dominant discourse surrounding contributions to science remains heteronormative, patriarchal, Western, and centred on the laboratory and academe. Though spaces for discourse exist for queer scientists (such as 500 Queer Scientists and Pride in STEM) and for Filipino scientists (such as Pinoy Scientists and ScienceChatPH), platforms for discussion and representation at the intersection of these identities are largely absent.

Queer Scientists PH was established to address these gaps through an online visibility campaign meant to empower younger STEM workers and attract a community of queer Filipino STEM workers globally. The effort seeks to challenge the dominant narratives of scientific contribution by having contributors talk about the challenges they have faced in their respective spaces, the successes they have experienced, and the hopes they hold onto for the LGBTIQA+ community and the scientific community.

Though many visibility campaigns such as 500 Queer Scientists and Pinoy Scientists revolve around individual self-submitted bios, the depersonalized effort leaves individual stories represented without fostering a sense of community and long-term connection to a larger network. We have created an open call for contributors through Twitter (Queer Scientists PH, 2022) and Instagram (Queer Scientists PH, 2021b) to encourage Filipino STEM workers globally to share their stories via an interview instead (Queer Scientists PH, 2021a).

Considering that being out online does not equate being out offline, contributors are informed beforehand about the goals of the organization and the social media platforms where their stories will be published – including a clause that enables them to withdraw their stories at any point, for any reason. Prior to publication on *Medium* (Queer Scientists PH, 2021c), the full transcript of the interview is returned to the contributors for approval to ensure information they disclosed can be publicized. Any information not publicized is treated with utmost confidentiality. Photos of contributors are self-submitted and are a mix of their personal and professional lives to humanize our contributors and decentralize the lab-dominated narratives of science. Features have also been carefully arranged to spotlight individuals from different scientific disciplines (from physics and bioengineering to neuroethics and science communication), geographical locations (from provinces in the Philippines to US), and SOGIEs (beyond strict binaries or labels and towards representing each person's process of self-discovery and self-expression). Feedback forms have been sent to contributors to evaluate our processes and their experiences participating in it. Contributor evaluations have been positive, and all respondents so far have also signified interest in participating in the campaign's future activities as mentors and speakers.

After almost a year on, the Twitter and Instagram pages of the campaign have developed a modest but dedicated following of 669 and 369 followers, respectively. Impressions on Twitter and Instagram peaked in June 2021 when three queer scientists were featured for three consecutive weeks with 214,000 and 4,100 impressions respectively. As of October 2021, the organization has grown from its initial team of three early-career researchers to a team of 20 volunteers, including non-science workers. We have also been featured in multiple news outlets and magazines, such as *Philippine Daily Inquirer*, *CNN Philippines Life*, *Scientia*, *Attitude Magazine*, and have partnered with several organizations and institutions, such as STEM+PH, Siyénsiyá PH, University of the Philippines Diliman College of Science, and Sanggunian: Commission on Gender Equality.

Current efforts towards improving the organization are centred on expanding the visibility campaign to represent not only regional members of the community (particularly those outside the country's capital Metro Manila, for example from Visayas and Mindanao), but also Filipino STEM workers who remain underrepresented within the LGBTIQA+ community. Community-centred and discussion-based initiatives such as talks and forums regarding pressing issues for these intersecting identities are being planned. The absence of an anti-discrimination bill in the Philippines also makes it necessary to hold SOGIE workshops for new volunteers and members, whether or not they belong to the LGBTIQA+ community. More community building activities are planned to help queer Filipino STEM students and early career researchers link with mentors who are members and/or allies of the LGBTIQA+ community and find opportunities for funding research, further studies here in the Philippines and abroad, and career alternatives. These efforts are geared not only towards educating those in and out of the community but also towards lobbying for systemic changes through connecting to collective social movements, creating position statements, and releasing public calls to action for LGBTQIA+ related issues.

Queer Scientists PH has evolved into an advocacy organization that supports the creation of a supportive and discrimination-free environment for LGBTIQA+ individuals in STEM and STEM-allied professions in the Philippines and around the world.

References

Queer Scientists PH. (2021a) 'Queer Scientists PH', Google form [online], January, Available from: https://tinyurl.com/queerscientistsph [Accessed 1 April 2022].

Queer Scientists PH. (2021b) 'Queer Scientists PH', *Instagram* [online], January, Available from: https://www.instagram.com/queerscienceph/ [Accessed 26 March 2022].

Queer Scientists PH. (2021c) 'Queer Scientists PH', *Medium* [Blog], September, Available from: https://queerscientistsph.medium.com/ [Accessed 1 April 2022].

Queer Scientists PH. (2022) 'Queer Scientists PH', *Twitter* [online] 11 February, Available from: https://twitter.com/queerscienceph [Accessed 26 March 2022].

Practice Spotlight: 500 Queer Scientists at the Sydney Gay and Lesbian Mardi Gras

Alice Motion and Hervé Sauquet

In 2018 two US-based scientists, Dr Lauren Esposito and Sean Vidal Edgerton, took the opportunity of Pride Month to launch the online visibility campaign, *500 Queer Scientists* (500 Queer Scientists, 2022). Inspired by this initiative, which shares portrait photographs and short biographical stories from queer scientists, we have run an annual event in Sydney since 2019, 'An Evening with 500 Queer Scientists'. Designed to celebrate Australia-based queer scientists and their research, the event was timed to coincide with the Sydney Gay and Lesbian Mardi Gras and held at the Royal Botanic Garden Sydney.

Building on the format of the *500 Queer Scientists* webpage, our event features a panel of five LGBTIQA+ scientists, each invited to speak for ten minutes about their life in science and to reflect on the ways in which their career intersects with their queer identity. A queer scientist emcees the event, weaving the individual stories together and facilitating discussion among the panellists before inviting questions from the audience and closing the event. The event is not recorded and its ephemeral nature generates a sense of realness and intimacy for those in attendance (about 140 people). Whatever happens will only be recalled in the faltering light of human memory or preserved in secondary retelling. Queer stories generate more queer stories and so community is built.

A lack of queer visibility in science is an ongoing challenge (Yoder and Mattheis, 2016) and this is further compounded by the fact that queer festivals, in common with other cultural festivals, have not historically featured science or science communication prominently. Our event has provided an important opportunity to place queer scientists in the spotlight as role models for the science, technology, engineering, mathematics, and medicine (STEMM) community, and to also establish science as part of our shared culture for Mardi Gras.

Audience feedback has been overwhelmingly positive, particularly from those who have struggled as closeted scientists at work. Additionally, during intervals and refreshment breaks, we have heard from attendees who have welcomed the expansion of the Mardi Gras programme to include scientific content. These members of the community have not felt catered to as part of Mardi Gras and told us that their identities as scientists and queer people were, in some ways, separate because of a lack of events at the intersection.

We have tried to ensure that our event is inclusive of people from different scientific disciplines, ages, disabilities, genders, queer identities, and cultural backgrounds, and to provide a safe space for people to share personal stories that are sometimes funny, sometimes sad, and always powerful (Clugston, 2021).

'An Evening with 500 Queer Scientists' served as the catalyst for the development of the New South Wales chapter of QueersInScience, a national organization aimed at building community and improving support for LGBTIQA+ people in STEMM in Australia.

In 2021, we reached an important milestone when our event was included in the official Mardi Gras Programme for the first time. The same year we were delighted to receive unsolicited feedback from Sue Fletcher, who in 1978 had been present at the gay and lesbian protest that Mardi Gras has commemorated every year since. Sue wrote, 'I think your event at the Royal Botanic Garden was one of the best Mardi Gras events I have been to in many years – the format is fantastic.'

We are proud of the success of this event and, more importantly, the community that we have created through the stories shared.

References

500 Queer Scientists. (2022) 'Stories', *500 Queer Scientists* [online], Available from: https://500queerscientists.com/ [Accessed 1 April 2022].

Clugston, J. (2021) 'An Evening with 500 Queer Scientists (26 Feb 2021)', *YouTube* [online] 7 May, Available from: https://youtube/8UosGn-LDeA [Accessed 11 November 2021].

Yoder, J.B. and Mattheis, A. (2016) 'Queer in STEM: workplace experiences reported in a national survey of LGBTQA individuals in science technology, engineering, and mathematics careers', *Journal of Homosexuality*, 63(1): 1–27.

7

Including Queerness and Improving Belonging of Intersectional Queer Identities in Science Communication Communities

Katherine Canfield

"Are there any other identities that are important to you that I haven't asked you about today?" I asked.
"I'm a lesbian," she declared with a soft smile, which I involuntarily reciprocated.

I felt a sense of simultaneous calm and excitement overwhelm as those words hung in the air and we closed out the video interview. While past interviewees had shared their LGBTIQA+ identities when asked this question, this time it resonated with me differently, reinforcing my sense of self and queer belonging in science communication. Researching efforts at building inclusive science communication (ISC) research and practice revealed that many of the leaders in this movement had numerous marginalized identities of which they live at the intersection. The ISC movement was a space where they both passively and actively were their full selves, in research, practice, and – simply – life.

This conversation exemplified a larger trend in the strengths and struggles of existing efforts at addressing inclusion in science communication and science, technology, engineering, and mathematics (STEM). Increasingly, organizations and lab groups are taking a holistic approach to equitable science communication and addressing common exclusionary practices (Judd and McKinnon, 2021). Despite this notable effort to meaningfully support inclusion among some axes of identity, a remaining challenge is creating science communication communities that are radically and intentionally inclusive of queerness. This chapter seeks to briefly present how this came to be and provide recommendations for how we might build science communication communities that are inclusive of intersectional queer identities moving forward. It focuses specifically on mobilizing communities in alignment with the importance of shared experiences and identity, relationship building, and self-efficacy to create social change (Orthia et al, 2021).

In this chapter, references to queerness may encompass aspects of sexuality, gender, and variations in sex characteristics, recognizing that these are distinct elements of identity, and yet also how the queer community aspires to include all people who have identities that are not binary-cisgender and/or heterosexual (Yoder and Mattheis, 2016).

Queerness in STEM and science communication communities

The STEM fields have been making nominal efforts towards equity for 50 years. Early conversations focused specifically on race and ethnic diversity in the STEM fields (Gillette, 1972) and – later – binary gender inequality without addressing queerness (Hanson et al, 1996; Settles et al, 2006). In scientific societies as well, while there are exceptions (Atherton et al, 2016), the conversation has remained largely about binary gender inclusion and inequity rather than focusing on the full spectrum of genders and other manifestations of queerness that are largely underrepresented and under-supported in STEM societies (Cumming-Potvin and Martino, 2018).

Since 2016, STEM researchers and organizations have increased their calls on the need for radical inclusion and support of queerness in these fields (Atherton et al, 2016; Scarpelli, 2017; Brusman, 2020). There are plenty of challenges for these calls to tackle. Queer scientists and communicators continually deal with 'status management', constantly having to decide how much of their queerness to share as they move through life due to the way this identity can be discriminated against and devalued (Cech and Rothwell, 2020). Further, LGBTIQA+ students are less likely to pursue tertiary studies in STEM than their heterosexual peers (Hughes, 2018). This aligns with findings that in academic research opportunities, especially in STEM, queer-identified people continue to experience isolation, exclusion, harassment, and less opportunities for advancement compared to their binary-cisgender and heterosexual peers (Yoder and Mattheis, 2016; Cech and Waidzunas, 2021). Outside of university settings, queer people in federal agencies in the US and no doubt equivalent bodies elsewhere also face notable exclusion and harassment (Cech and Pham, 2017; Cech and Rothwell, 2020).

In science communication, recent conversations have considered advancing diversity in the field, largely still focused on the axes of binary gender and race, with lesser considerations of disability and other axes of difference (Massarani and Merzagora, 2014; Rasekoala, 2019; Judd and McKinnon, 2021). Beyond this, there are isolated case studies exploring non-binary, gender diverse, and transgender experiences (Barres, 2006; Pérez-Bustos, 2014). The relative dearth of academic documentation of the queer experience in science communication, and especially as related to intersectional identities (Crenshaw, 1989), invites further investigation and identification of promising practices to promote radical inclusion of queerness in science communication (Judd and McKinnon, 2021; Roberson and Orthia, 2021).

Queerness and community

There are many different forms of communities. For the purposes of this chapter, I define community using an interpretive approach (Franz et al, 2018), where the community is defined through the lived experience of members who relate from having a shared sense of identity (Edberg et al, 2015).

The importance of community for queer people has been explored significantly through both peer-reviewed (d'Emilio, 1983) and popular literature (Brammer, 2021). This literature has emphasized the way community among queer people allows for creation of a 'chosen family', which serves to create a sense of belonging and safety for queer people. This is particularly the case among people whose queerness was either passively or actively rejected by the families with whom they grew up, to find a sense of home outside of blood relations. Further, rather than there being a singular, homogenous LGBTIQA+ community, there are multiple queer publics reflecting different shared identities (Ciszek, 2020). People who have more privileged identities (for example, a white cisgender gay man) have different experiences than people with other combinations of race, gender, and sexuality identities, both within science communication and in society writ large (Bowleg, 2008).

Queering science communication communities is more than a goal of acknowledging and accepting LGBTIQA+ people in the space (Feng, 2021). Queering challenges the power relations of heteronormativity through direct actions that centre queer experiences (Roberson and Orthia, 2021) with the goal of achieving justice and equity for queer people (Abu-Assab and Nasser-Edin, 2018; Feng, 2021).

Through a variety of online and in person communities, science communication has begun to address lasting exclusion of marginalized identities from science communication, with room for significant growth on the queering front. These communities exist in Twitter hashtags and accounts (Reed, 2013; Canfield et al, 2020), non-profits (Bert, 2019; Roca et al, 2020), and affinity groups (Valdez-Ward et al, 2019). Recent efforts to document and improve inclusion of marginalized identities in STEM (Bevan et al, 2018) and science communication (Canfield and Menezes, 2020; YESTEM Project UK Team, 2020) further define these communities. While these efforts have not specifically focused on how queer community members' intersectional identities are supported, past efforts at building inclusive spaces across other identities in science communication provide hope for progress. For example, the US-based ISC movement is built on researchers and communicators bringing their full selves to all that they do and operating with intentionality, reciprocity, and reflexivity, but to date this has been largely explored at a holistic rather than community-specific scale (Canfield and Menezes, 2020).

Poor practices to avoid

Effectively queering science communication communities will require thoughtful implementation of recommended practices and avoiding solutions that perpetuate exclusion of the many axes of identity that intersect with queerness. Existing research that has identified barriers to inclusive science communication can inform the more specific experiences of queerness. Importantly, in identifying these practices, it is not only the identity of queerness that needs to be considered, but the way queerness interacts with other identities to create experiences of inclusion or exclusion in science communication.

One major practice to avoid is emphasizing 'one size fits all' solutions. Prescriptive solutions are attractive in that they promise learning a few strategies well could solve many problems, but such approaches erase nuanced needs and context. Past research on inclusive design for science communication has emphasized the importance of learning the specific needs of each community with whom you work (Dawson, 2019; Canfield and Menezes, 2020). As an example, it would be potentially harmful to assume that practices used to include cisgender, able-bodied queer people with postgraduate degrees and employment would be equally effective in reaching queer undergraduate students considering careers in science communication with a diversity of disability status and gender identities. The needs of these groups would differ on these different axes of identity (Grillo, 1995).

Assuming is closely tied to the dangers of prescriptive solutions, where applying one solution to many communities and scenarios assumes that the needs are the same. Especially in addressing inclusion of intersectional identities, it is harmful to assume that visual cues defined by our expected presentations of ability, gender, sexuality, and other identities are aligned with people's identities (Grillo, 1995; Bennett, 2000). Making a thorough effort to understand the identities of community members who have been excluded – while also being thoughtful of giving people an opportunity rather than requirement to share their identities – will help to develop more nuanced practices that address needs more effectively (Cooper et al, 2020).

Further, assumptions should also be avoided in planning efforts, to improve intersectional inclusion of queerness in science communication. This includes assuming that cisgender and heterosexual people in a science communication group, or people who are otherwise not fully aligned with the identities of the group you are working to include, are able to accurately define the goals of an inclusive effort. It should also not be assumed that cis-heterosexual people will make appropriate decisions on (1) terminology related to queerness, (2) who the relevant players are in the community, and (3) what the appropriate process or procedures are for building out an inclusive effort. It is essential that these decisions are all made in close coordination with the communities that have been excluded to ensure their actual experiences are informing the process and their needs are being addressed (Chilvers, 2013).

Relatedly, while broader existing research on challenges to achieving inclusive science communication communities can help define some challenges of being queer in science communication, it is also important to understand how queerness has its own unique challenges. Practices that are recommended for addressing inclusion along other axes of identity may not be perfectly transferable to address the exclusions of queerness. Especially since queerness is not always a visible identity, recommendations addressing discrimination based on appearance will not tackle all the challenges of queerness, or of those people who choose to keep their queerness more unseen. Recommended practices from other studies may be helpful scaffolding in designing practices for queering science communication communities when accompanied by further practices to thoughtfully address challenges specific to queerness and queer exclusion.

A final practice to avoid is designing purely single-issue solutions. As Audre Lorde put it, 'there is no such thing as a single-issue struggle, because we do not live single-issue lives' (2007, p 138). As a queer, Black woman, she is an historic example of how creating single issue solutions to challenges of queerness will not be sufficient to tackle the challenges and injustices faced in our nuanced, intersectional lives. While there may certainly be situations in which it is appropriate to design smaller efforts that address individual aspects of queer exclusion at a time, it is also important to design efforts that address challenges across multiple identities. One example of a single-issue solution failing to meet intersectional needs in the past was in having gender inclusive restrooms marked at the inaugural #InclusiveSciComm Symposium, but in a relatively inaccessible location for disabled participants.[1]

A holistic plan for inclusion may map out the many small efforts that make up the overall initiative and how they interact to address the many issues at play in creating more inclusive science communication communities for intersectional queer identities.

Steps towards radical inclusion

The conversations around improving sense of belonging and inclusion for identities underrepresented in STEM and science communication today are trying to address many of the same challenges as when they began fifty years ago, pointing to the need for a new approach in these efforts. As we move through an extended global moment of people waking up to privilege, there has been a renewed desire to learn how to build more inclusive, and specifically antiracist, fields of STEM and science communication. In so doing, it is important to recognize how building an antiracist space also necessitates an intersectional lens (Crenshaw, 1989) that addresses intersectional marginalization (Strolovitch, 2007) and focuses on radical inclusion of queer identities.

Some scholars are explicitly working to tackle intersectional marginalizations through peer reviewed literature and terminology. Queer theory has a history of being largely white-centric (Johnson, 2001). Reimagining terminology to recognize the intersection of racial and ethnic identities with queerness (Glick, 2003; Hames-García, 2013) and the intersections of critical race and queer theory (Ciszek, 2018; Johnson, 2001) will provide theoretically grounded opportunities to advocate for a queerer academic space. Further, in advocating for intersectional queer identities, intentionally defining science and science communication rather than defaulting to a white, Western, and ableist definition of the fields and terms can help promote a greater sense of belonging in the field (Finlay et al, 2021). While not all science communication communities are within the confines of academia, having theoretical arguments for tackling intersectional marginalization in academic science communication communities supports advocacy in these spaces. Grounding of academic theory can provide leverage in advocating for institutional scale change at academic institutions rather than the less effective piecemeal or short-term efforts (Dawson, 2014).

Another recommended practice is to build inclusion of queerness into the institutions and communities themselves. Rather than having a single diversity committee that is tasked with all efforts related to inclusion, building goals of radical inclusion and justice into groups' charters and missions is one way to make this goal central to all members' work. The American Physical Society created a code of conduct that included considerations for LGBTIQA+ physicists as one mechanism for centring consideration of the queer communities and their needs within the organization (Atherton et al, 2016). Along with creating charters that provide specific expectations of what just inclusion of intersectional queer identities looks like, institutional change requires communities to reflect on and analyse the overall culture of their group. Collecting perspectives of all group members can provide input on how to improve the organization. Using multiple methods of evaluation can ensure there are diverse opportunities for people to contribute in a democratic way and will allow for more thorough evaluation (Bandy et al, 2018). This may reveal simple shifts that can be made to be more inclusive regarding the language used to refer to LGBTIQA+ people, and identify other potentially exclusive practices (for example, limited gender identity options on conference or group registration) that can be shifted to create a better experience for queer members (Cooper et al, 2020).

Along with the importance of institutional scale change, radical inclusion of queer identity also requires reflexivity at the individual scale. This is one of the key tenets of ISC and is a call for individuals to continually and critically reflect on their own identities, practices, and outcomes to ensure they are always being equitable, and to adapt their approach as needed (Canfield and Menezes, 2020). This reflexivity requires individuals to also be aware of how their identities are reflective, or not, of the communities with whom they work. If such reflection reveals a great misalignment in identity or values between the science communicator and the audience or community, it is likely an inappropriate space for that communicator to work in. In queering science communication communities, this is a call for cis-heterosexual leaders to step back and make sure there is queer leadership on their teams, as the lived experiences of queer people provides essential expertise to building a just environment across identities.

Science communication leaders' embodiment of their intersectional identities has promoted and can continue to promote a sense of belonging for me and other members of the queer community in science communication and society (Canfield et al, 2020; Roberson and Orthia, 2021). Along with the queerness that some science communicators boldly own, others bring their full identities to their work along the axes of ethnicity (Aguado and Porras, 2020), immigration status (Valdez-Ward, 2018), and more. In conducting my own work on ISC, the research process revealed the importance of reflexivity among the leaders and led me to reflect on my own identities. The confidence with which these leaders fully occupied their identities helped provide me with a great sense of belonging in this field and built my confidence that I can proudly wear and own my queer and other identities as a science communication researcher (Chilvers, 2013). Thus, another opportunity for radical inclusion of queerness in science communication is in continuing to create spaces where people can proudly be themselves. This requires intentional definition of who is included in communities, so that queer science communicators can feel as safe as possible to be themselves.

An overall paradigm shift towards intersectional inclusion and queer experience and belonging has great potential to improve support for queer people both in science communication and STEM fields. This will require deconstructing power structures, a central goal of queer theory (Glick, 2003), and inclusive science communication (Canfield and Menezes, 2020). Such a goal requires us to ask not 'what is?' but rather 'what can be?' (Gunkel, 2009). Science communication can be a space inclusive of intersectional queer identities. Using the practices discussed here, we can begin to increase representation while radicalizing approaches to move towards the norm being that people can feel confident in the queer inclusiveness of science communication.

Note

[1] This is not a published finding, but was discussed in a 'lessons learned' final session of the 2018 #InclusiveSciComm Symposium held at the University of Rhode Island, Kingston, RI, USA, 29 September.

References

Abu-Assab, N. and Nasser-Eddin, N. (2018) 'Queering justice: states as machines of oppression', *Kohl: A Journal for Body and Gender Research*, 4(1): 48–59.

Aguado, B.A. and Porras, A.M. (2020) 'Building a virtual community to support and celebrate the success of Latinx scientists', *Nature Reviews Materials*, 5(12): 862–864.

Atherton, T.J., Barthelemy, R.S., Deconinck, W., Falk, M.L., Garmon, S., Long, et al. (2016) *LGBT Climate in Physics: Building an Inclusive Community*, College Park: American Physical Society.

Bandy, J., Price, M.F., Clayton, P.H., Metzker, J., Nigro, G., Stanlick, S., et al. (2018) *Democratically Engaged Assessment: Reimagining the Purposes and Practices of Assessment in Community Engagement*, Davis: Imagining America.

Barres, B.A. (2006) 'Does gender matter?', *Nature*, 442(7099): 133–136.

Bennett, J.M. (2000) '"Lesbian-like" and the social history of lesbianisms', *Journal of the History of Sexuality*, 9(1/2): 1–24.

Bert, A. (2019) 'On being LGBTQ+ in science – yes it matters, and here's why', *Elsevier Connect* [online] 25 July, Available from: https://www.elsevier.com/connect/on-being-lgbtq-in-science-yes-it-matters-and-heres-why [Accessed 1 October 2021].

Bevan, B., Calabrese Barton, A., and Garibay, C. (2018) *Broadening Perspectives on Broadening Participation in STEM*, Washington, DC: CAISE.

Bowleg, L. 2008. 'When Black + Lesbian + Woman ≠ Black Lesbian Woman: the methodological challenges to qualitative and quantitative intersectionality research', *Sex Roles*, 59: 312–325.

Brammer, J. (2021) 'The queer comforts of community', *Harper's Bazaar* [online] 2 June, Available from: https://www.harpersbazaar.com/culture/features/a36620381/jp-brammer-queer-comforts-community-essay/ [Accessed 4 June 2021].

Brusman, L. (2020) 'Science thinks it's unbiased. Queer scientists know that's untrue', *Massive Science* [online] 18 June, Available from: https://massivesci.com/articles/queer-science-stem-lgbtqia-bias-field-work-lab-universities-school-discrimination/ [Accessed 5 Oct 2021].

Canfield, K.N., Menezes, S., Matsuda, S.B., Moore, A., Mosley Austin, A.N., Dewsbury, B.M., et al. (2020) 'Science communication demands a critical approach that centers inclusion, equity, and intersectionality', *Frontiers in Communication*, 5: 2.

Canfield, K.N. and Menezes, S. (2020) *The State of Inclusive Science Communication: A Landscape Study*, Kingston: Metcalf Institute.

Cech, E.A. and Pham, M.V. (2017) 'Queer in STEM organizations: workplace disadvantages for LGBT employees in STEM related federal agencies', *Social Sciences*, 6(1): 12.

Cech, E.A. and Rothwell, W.R. (2020) 'LGBT workplace inequality in the federal workforce: intersectional processes, organizational contexts, and turnover considerations', *ILR Review*, 73(1): 25–60.

Cech, E.A. and Waidzunas, T.J. (2021) 'Systemic inequalities for LGBTQ professionals in STEM', *Science Advances*, 7(3): eabe0933.

Chilvers, J. (2013) 'Reflexive engagement? Actors, learning, and reflexivity in public dialogue on science and technology', *Science Communication*, 35(3): 283–310.

Ciszek, E. (2018) 'Queering PR: directions in theory and research for public relations scholarship', *Journal of Public Relations Research*, 30(4): 134–145.

Ciszek, E. (2020) '"We are people, not transactions": trust as a precursor to dialogue with LGBTQ publics', *Public Relations Review*, 46(1): 101759.

Cooper, K.M., Auerbach, A.J.J., Bader, J.D., Beadles-Bohling, A.S., Brashears, J.A., Cline, E., ... and Brownell, S.E. (2020) 'Fourteen recommendations to create a more inclusive environment for LGBTQ+ individuals in academic biology', *CBE – Life Sciences Education*, 19(3): es6.

Crenshaw, K. (1989) 'Demarginalizing the intersection of race and sex: a Black feminist critique of antidiscrimination doctrine, feminist theory and antiracist politics', *University of Chicago Legal Forum*, 139: 139–168.

Cumming-Potvin, W.M. and Martino, W. (2018) 'Countering heteronormativity and cisnormativity in Australian schools: examining English teachers' reflections on gender and sexual diversity in the classroom', *Teaching and Teacher Education*, 74: 35–48.

d'Emilio, J. (1983) 'Capitalism and gay identity', in A. Snitow, C. Stansell, and S. Thompson (eds) *Powers of Desire: The Politics of Sexuality*, New Feminist Library Series, New York: Monthly Review Press.

Dawson, E. (2014) 'Reframing social exclusion from science communication: moving away from 'barriers' towards a more complex perspective', *Journal of Science Communication*, 13(2): 1–5.

Dawson, E. (2019) *Equity, Exclusion and Everyday Science Learning: The Experiences of Minoritised Groups*, Oxon: Routledge.

Edberg, M., Cleary, S., Simmons, L., Cubilla-Batista, I., Andrade, E., and Gudger, G. (2015) 'Defining the "community": applying ethnographic methods for a Latino immigrant health intervention', *Human Organization*, 74(1): 27–41.

Feng, J. (2021) 'Coming out of the Closet; Coming out as Environmentalists', Paper presented at the *7th Annual Duck Family Graduate Workshop on Environmental Politics and Governance* [online] 13 May.

Finlay, S.M., Raman, S., Rasekoala, E., Mignan, V., Dawson, E., Neeley, L., and Orthia, L.A. (2021) 'From the margins to the mainstream: deconstructing science communication as a white, Western paradigm', *Journal of Science Communication*, 20(1): C02.

Franz, B.A., Skinner, D., and Murphy, J.W. (2018) 'Defining "Community" in community health evaluation: perspectives from a sample of nonprofit Appalachian hospitals', *American Journal of Evaluation*, 39(2): 237–256.

Gillette, R. (1972) 'Minorities in the geosciences: beyond the open door', *Science* 177(4044): 148–151.

Glick, E. (2003) 'Introduction: defining queer ethnicities', *GLQ: A Journal of Lesbian and Gay Studies*, 10(1): 123–124.

Grillo, T. (1995) 'Anti-essentialism and intersectionality: tools to dismantle the master's house', *Berkeley Women's Law Journal*, 10(1): 16–30.

Gunkel, H. (2009) 'Through the postcolonial eyes: images of gender and female sexuality in contemporary South Africa', *Journal of Lesbian Studies*, 13(1): 77–87.

Hames-García, M. (2013) 'What's after queer theory? Queer ethnic and indigenous studies', *Feminist Studies*, 39(2): 384–404.

Hanson, S.L., Schaub, M., and Baker, D.P. (1996) 'Gender stratification in the science pipeline: a comparative analysis of seven countries', *Gender & Society*, 10(3): 271–90.

Hughes, B.E. (2018) 'Coming out in STEM: factors affecting retention of sexual minority STEM students', *Science Advances*, 4(3): eaao6373.

Johnson, E.P. (2001) '"Quare" studies, or (almost) everything I know about queer studies I learned from my grandmother', *Text and Performance Quarterly*, 21(1): 1–25.

Judd, K. and McKinnon, M. (2021) 'A systematic map of inclusion, equity and diversity in science communication research: do we practice what we preach?', *Frontiers in Communication*, 6: 744365.

Lorde, A. (2007) 'Learning from the 60s', in *Sister Outsider: Essays & Speeches by Audre Lorde*. Berkeley: Crossing Press, pp 134–144.

Massarani, L. and Merzagora, M. (2014) 'Socially inclusive science communication', *Journal of Science Communication*, 13(2): C01.

Orthia, L.A., McKinnon, M., Viaña, J.N., and Walker, G.W. (2021) 'Reorienting science communication towards communities', *Journal of Science Communication*, 20(03): A12.

Pérez-Bustos, T. (2014) 'Of caring practices in the public communication of science: seeing through trans women scientists' experiences', *Signs: Journal of Women in Culture and Society*, 39(4): 857–866.

Rasekoala, E. (2019) 'The seeming paradox of the need for a feminist agenda for science communication and the notion of science communication as a "ghetto" of women's over-representation: perspectives, interrogations and nuances from the global south', *Journal of Science Communication*, 18(04): C07.

Reed, P. (2013) 'Hashtags and retweets: using Twitter to aid community, communication and casual (informal) learning', *Research in Learning Technology*, 21: n.p.

Roberson, T. and Orthia, L.A. (2021) 'Queer world-making: a need for integrated intersectionality in science communication', *Journal of Science Communication*, 20(01): C05.

Roca, A.I., Colman, C.L., Haelle, T.S., and Lee, D.N. (2020) 'The SciCommDiversity Travel Fellowship: the challenge of creating a sustainable intervention', *Frontiers in Communication*, 5: 51.

Scarpelli, A. (2017) 'How I realized that LGBT+ scientists like me can inspire others in their field', *Massive Science* [online] 19 October, Available from: https://massivesci.com/articles/stem-lgbt-diversity-science/ [Accessed 20 July 2021].

Settles, I.H., Cortina, L.M., Malley, J., and Stewart, A.J. (2006) 'The climate for women in academic science: the good, the bad, and the changeable', *Psychology of Women Quarterly*, 30(1): 47–58.

Strolovitch, D.Z. (2007) *Affirmative Advocacy: Race, Class, and Gender in Interest Group Politics*, Chicago: University of Chicago Press.

Valdez-Ward, E. (2018) 'The end of DACA would be a blow to science', *Scientific American* [online] 12 December, Available from: https://blogs.scientificamerican.com/voices/the-end-of-daca-would-be-a-blow-to-science/ [Accessed 1 October 2021].

Valdez-Ward, E., Ulrich, R.N., Marcus, T.S., and Cat, L.A. (2019) 'Reclaiming STEM: bringing your identity and culture to STEM', *AGU Fall Meeting Abstracts*, 2019: U33B-05.

YESTEM Project UK Team (2020) 'The equity compass: a tool for supporting socially just practice', *The YESTEM Project* [website], Available from: http://yestem.org/wp-content/uploads/2021/09/2021-YESTEM-Insight-1-Equity-Compass-for-ISL-updated-Sept-2021.pdf [Accessed 1 October 2021].

Yoder, J.B. and Mattheis, A. (2016) 'Queer in STEM: workplace experiences reported in a national survey of LGBTQA individuals in science, technology, engineering, and mathematics careers', *Journal of Homosexuality*, 63(1): 1–27.

Practice Spotlight: Rainbow Spectrums: Embracing Our Queer Disabled Family in Science Communication

V de Kauwe and Kai Fisher

In 2019, the disability group Science Alliance and local advocacy organization ACT Down Syndrome organized an Australian National Science Week event called 'Rainbow Spectrums'. The aim was to showcase successful scientists who were both on the disability spectrum and within the LGBTIQA+ rainbow. Although this targeted a niche audience, the event made a deep impact on many present. In this spotlight, the head of Science Alliance describes Rainbow Spectrums, while an audience member describes its outcomes in his life as a disabled queer trans man.[1]

The event

Rainbow Spectrums consisted of a panel of eight: one author, editor, and senior lecturer; four other academics; one high school student; and two disability educators. It began with the head of Science Alliance giving a presentation of the current lack of visibility when it comes to queer disabled scientists. This included an exposition of how the current education system, as well as popular culture, negatively impact both the identity and agency of people who are queer and/or disabled. The aim was not to indicate that queer disabled scientists were 'just like everyone else' neither was it the sentiment that they were scientists 'even though they were disabled'. Such notions have the underlying assumption that neurotypical, able-bodied cis-heteronormality is the measure that queer disabled scientists must meet. Rather, this presentation revealed that there are valid members of the science community who are queer and disabled. After this, each panellist spoke of their role in the science community, how they navigate this role with regard to their disability, and how they experience all of this as a queer person. This was followed by questions and comments from the audience. The remaining time was spent with the academics providing hands-on demonstrations of their field of expertise. This final stage was more casual, with the audience members freely walking between each demonstration. During this time, a safe sex stall was made available, so that audience members could receive personal advice, information booklets, condoms, dental dams, latex-free gloves, and other equipment.

The aim of this event was visibility. For this event, queer visibility and disability empowerment were inseparable goals, presented within the context of strengthening queer disabled identities within the science community. This being the case, it seems appropriate for an audience member to speak for themselves regarding the outcome of the event.

The outcome: introducing Kai Fisher

Hi, my name is Kai, I'm a queer trans man, I also have multiple disabilities, and I love science. I came to this event not knowing what to expect: you never see anything like this anywhere. You never see science acknowledging that you might be gay AND with disabilities. There are some science events for people with disabilities, but these are mainly for kids. These can be fun, but they don't show that people with disabilities can have their minds engaged too. Serious science for disabilities, usually means visible disabilities; but Rainbow Spectrum included invisible disabilities too. And I haven't seen anything that combines queers and science either. We start to see it now in science fiction, like in *Doctor Who*, but not out here in the real world. And never, NEVER, have I dreamed that there could be something for queerness and disability together. So I came along.

The result was shocking: good shocks and bad shocks. I knew the rest of the audience felt the same because of the gasps and cheers. The bad shock was how badly the education system has let us down, even so-called disability-inclusion systems. But it was good to finally hear it said out loud, with researched proof and everything.

The good shock was to see and hear the presenters. Their voices were disabled queer voices, but they were so confident and competent. They spoke of their identity and disabilities with the same power that they talked about science. It shows that disabled queers CAN have an open existence in the science world: an existence that acknowledges sexuality and disabilities, but without asking for fixing. None of us needs fixing.

One recommendation I would make for similar events is to screen your panel. At one point one of the educators said that she had two trans children and that she thought that being trans was part of their disability. My heart broke for a second. But two of the researchers corrected the educator and the captain of Science Alliance repeated what she said about identity. She basically said that each person is the greatest expert of their own experience, no matter what their level of abilities. Even a baby can cry: that's the baby saying, 'I am here and this is what I need'. That strong science comeback made me think, 'Yes, fuck you, I can do this: trans is not "broken" and disability is not "broken" either. The scientists said so.'

Note

[1] At the time of writing the acceptable language around disability included 'people with disabilities' or 'disabled'. The former is more commonly used in Australia, while the latter is preferred internationally. Both are used in the chapter with respect and solidarity. Both authors openly acknowledge their disabilities.

8

Have Rainbow, Will Collect Data: How Citizen and Community Science Engages Queer Volunteers

Todd A. Harwell

Introduction

The vast majority of science learning and engagement occurs beyond the walls of a classroom across the span of one's lifetime (Falk and Dierking, 2010). Given this fact, there is a large portion of the global population who seek opportunities that allow them to pursue their lifelong scientific interests and curiosities in a variety of settings. These opportunities can be considered informal science, technology, engineering, and mathematics (STEM) engagement opportunities and include people of all ages, interests, and backgrounds who engage with STEM on their own initiative through dynamic mediums. Among informal opportunities that promote public participation in STEM are those in which the public engages with and contributes to scientific research activities. These types of projects and programmes may include and/or be referred to as examples of volunteer monitoring, public participation in scientific research (PPRS), or more generally, as citizen and community science (CCS). Regardless of the name or title ascribed to this form of participatory STEM engagement, these opportunities largely remain rooted in traditional scientific norms and practices, which can often be exclusionary or unwelcoming for individuals with marginalized identities and backgrounds, including members of the LGBTIQA+ community.

In a broad sense, CCS can be used to identify an array of projects that involve 'nonscientists engaging in doing science' (NASEM, 2018). CCS describes activities that typically involve public audiences in the process, methods, and standards of scientific research with the aim of advancing scientific knowledge or application (NASEM, 2018). As a practice and a field, CCS is not well defined or well bounded, however, it is becoming the term most commonly recognized within the scientific and science communication communities. Beyond engaging in doing science, which commonly involves the processes of data collection and scientific research, CCS engagement serves as an example of informal science learning by engaging volunteers with scientific content and processes (Bonney et al, 2009). While the responsibilities and duties of participants in CCS may vary based on a variety of project design factors and overall project model of participation (Shirk et al, 2012), generally the role of a CCS volunteer is to assist with and contribute to scientific research or monitoring activities.

As CCS opportunities have become more prevalent across STEM research and engagement, there has also been an increase in research about the perspectives and experiences of participants. Within the body of research on CCS practice there has been a shifting focus to examine the motivations, contributions, and general experiences of the individuals who engage with these opportunities – CCS volunteers. More recently, there have been increased efforts to acknowledge and understand the individual and collective complexities of those who are gaining knowledge and contributing to these scientific outcomes. CCS opportunities allow individuals and communities to deepen their connections to science by promoting both personal and shared science identities (Farland-Smith, 2012; NASEM, 2018; Rahm and Ash, 2008), especially for participants from communities and backgrounds considered to be nondominant and underrepresented in STEM (NASEM, 2018). Researchers have also found that elements related to the social dimensions of engagement, including a sense of belonging to a community of volunteers, has contributed to volunteers' level of participation (Jennett et al, 2016) as well as their motivations to continue to engage with their projects (Aristeidou et al, 2015). Along the same lines, the development and fostering of social networks and communities has been found to play an important role in contributing to citizen science volunteer satisfaction and participation (Haywood, 2016) as well as to the general health and wellbeing of individuals from non-dominant communities and identities, especially those who identify as LGBTIQA+ (Frost and Meyer, 2012).

While there have been previous studies regarding the engagement of underrepresented CCS volunteers based on race/ethnicity, (binary) sex/gender, and/or socioeconomic status (Cooper and Smith, 2010; Soleri et al, 2016; Ganzevoort, et al, 2017; Dawson, 2018; Sorensen et al, 2019; Davis et al, 2020), little attention is given to the engagement of LGBTIQA+ individuals who participate in CCS. Yet, it is important for CCS practitioners and researchers to understand the unique ways that LGBTIQA+ individuals not only experience the world but perceive and approach science, science communication, and science engagement in order to continue to diversify and broaden the global STEM workforce while also recognizing, valuing, and embracing queer approaches to science.

CCS has been recognized as one of the most impactful developments in science communication of the last generation (Lewenstein, 2016) and as such it has become an important avenue to explore as it provides a multitude of opportunities for lifelong STEM engagement. This chapter will highlight (the few) examples of CCS projects that focus on queer topics, issues, or participants, and share an exploration of the experiences of LGBTIQA+ identifying CCS volunteers. It aims to illuminate whether CCS opportunities may be a mechanism for broadening the participation of LGBTIQA+ individuals in STEM via their engagement as CCS volunteers. By looking at the complex relationships between science and queer identities, as well as understanding the nature of community regarding their CCS project volunteers and the broader queer community, we can begin to understand how to leverage CCS as a means of queering science and increasing LGBTIQA+ engagement in STEM in a welcoming, safe, and inclusive way.

Positionality

In scholarly work, and in life in general, a researcher's relevant social positioning of being an 'insider' and sharing experiences of the communities and individuals that are of interest to them contributes to the overall reflexivity as a benefit to the participants' experiences and overall scholarship (Berger, 2015). As part of the reflexive process, it is vital to acknowledge the common and distinct identities, experiences, and interests of scholars and participants alike. In this regard, it is important to note that I identify as a gay/queer white cisgender man who is in a marriage and partnership with a gay/queer Hispanic (of South American origin) cisgender man. Additionally, I have engaged in citizen and community science both as a volunteer and as an instructive tool in my role as educator of STEM undergraduate courses. My primary research interests surround exploring the relationships between informal STEM engagement opportunities and the LGBTIQA+ community.

Queer science topics and issues in citizen and community science

What better way to study queer science topics than to tap into a broad network of folks who are scientifically curious and also identify as LGBTIQA+? It would appear that leveraging citizen and community science practices to collect data about queer topics directly from queer folks would be an obvious methodological approach. However, it seems that researchers have not had these same inclinations. While there is impactful work and scholarship being done surrounding queer topics and issues, often by researchers who self-identify as LGBTIQA+ and utilize queer theoretical perspectives, queer methodologies, and engage queer participants, to date there have been virtually no examples of an explicitly queer citizen or community science approach to investigating LGBTIQA+ topics or issues.

Despite my efforts to comb through global project databases, networks, and a broad internet search over the course of two years as part of my broader dissertation research, only one project emerged as having any sort of explicit queer focus. This project, based in Anchorage, Alaska in the United States, focuses on promoting healthy aging and community-wide respect and inclusion for LGBT elders in Anchorage. Led by Dr Diane King, the project aims to assess the feasibility of using *Our Voice*, a participatory electronic data collection tool that captures photo and narrative data, as a citizen science method focused on promoting safe, respectful, and inclusive community environments based on the understanding that LGBT elders can experience health disparities stemming from social stigma, loneliness, social isolation, and discrimination (King et al, 2020). Leveraging partnerships with the Alaskan chapters of Services and Advocacy for LGBT Elders (SAGE) and the American Association of Retired Persons (AARP), this method involves engaging LGBT individuals over the age of 50 in using an app to take geo-coded photos along with recordings to highlight environmental features that enabled or hindered safe and healthy aging. Some of the general findings revealed that lack of public transportation and information about LGBT-friendly places and services hindered the participants' personal safety, respect, inclusion, social participation, and connectedness (King et al, 2020).

Engaging queer citizen and community science volunteers

Beyond queer topics and issues within CCS, the engagement of LGBTIQA+ volunteers and contributors seems to be among the primary connections between the queer community and lifelong informal STEM engagement. The field of CCS and broader scientific community alike should look to the CCS volunteers directly to more fully understand how aspects of their backgrounds, lived experiences, and relationships with others impact their individual outcomes and project contributions, and how they influence their relationships with science and fellow volunteers. In order to better understand the queer CCS experience, I spoke with 14 LGBTIQA+ CCS volunteers from the United States to get a glimpse into how their queerness, as it related to their involvement in CCS, impacted their relationships with science and with fellow volunteers.

Personal perspectives and experiences of queer volunteers

It is no secret that LGBTIQA+ individuals experience life and the world around them uniquely as shaped by facets related to their queer identities. Their queer perspectives directly influence their relationships with science along with their engagement in CCS, albeit perhaps in ways that may be subconscious. Many of the queer CCS volunteers I interviewed observed that the field of CCS and its breadth of opportunities to informally engage with science seem to attract a greater proportion of LGBTIQA+ folks compared to more formal academic and professional STEM environments. While there is little to no demographic data available to back up their observations, given that sexual orientation and gender identities beyond the binary are seldom tracked in academic, professional, and CCS organizations alike, anecdotally these LGBTIQA+ CCS volunteers believed that there is a decent representation of queer folks engaged in CCS. A 39-year-old white, gay/queer man noticed that there were many LGBTIQA+ individuals that were connected to organizations and projects overlapping with his CCS affiliations. He reflected:

> One observation is that a lot of folks that do data intensive environmental work in [this city] are queer, or, on the rainbow. It's often sort of an inside joke of like the green queer mafia in [the city], so that pops up and a lot of volunteers that have come through over the years – they identified as LGBTQ.

Some of the queer CCS volunteers also conveyed that CCS opportunities offer a more welcoming atmosphere compared to formal STEM hubs, which are especially appealing to queer folks. They reflected that CCS was a safer space to authentically engage with science compared to a formal laboratory setting found at a university or within industry. A 23-year-old white, queer/genderqueer individual described how they viewed CCS as a bridge between the queer community and the scientific community. They shared:

> Citizen science gave me a chance to, like, no longer worry that there isn't a scientist that looks like me, or has the same gender or sexuality as me. But right now, I am that scientist doing that research, and so it doesn't matter because I'm that scientist in that moment ... and so I think citizen science was that new space where you could do science but also be out in yourself at the same time with a group of people who are like that.

In terms of authentically engaging with science via CCS, a number of LGBTIQA+ volunteers acknowledged that not only do personal identities impact how an individual pursues scientific activities, but queer identities are an asset in science. These individuals described ways that aspects of their identities as a queer person contribute to how they view and contribute to science. A 29-year-old white, queer, and transgender/non-binary individual expressed how individuals with marginalized identities approach the world quite differently than individuals from dominant cultures, communities, and identities. They stated:

> Being queer or being a person of color or being any number of things can give you life experiences that leads you to ask different questions or to have different insights when you have a problem ... science actually requires a lot of creativity like coming up with research questions and figuring out how you're going to answer those research questions and what kind of data will really answer it and how you will get that data ... so I think I can offer new insights just on the basis of being different ... if you're marginalized in some way.

These sentiments were echoed by a 36-year-old white, queer/asexual/demisexual and agender/non-binary volunteer. They also supported the notion that one's queerness, along with other marginalized or non-dominant aspects of identity such as ableness, contribute to diversifying scientific approaches. They commented:

> I definitely know that there's a lot of perspectives that queer people have that the mainstream population doesn't have, and we can bring a lot of innovation to the field – that's the case with any level of diversity. I'm also disabled, and there's a lot of shortcuts I can take and ways that I work as efficiently as possible that are a really big factor in my success in science. So I think anytime you have more diversity, the group is going to be better and it's going to make more contributions.

Queer volunteers lacking a sense of community

Along with the personal dimensions of engagement in CCS, a variety of social factors play an important role in a queer volunteer's overall engagement and contributions to scientific objectives and activities. When asked if they would describe the group of volunteers that engage with their CCS projects as a community, many queer volunteers responded with a simple "no" or "not really". For this group of individuals, there is great potential to foster a sense of community among CCS volunteers, however, this is rarely the case in their experiences. The lack of a sense of community can be considered a barrier to sustained engagement in CCS, and was commonly reported by many of the 14 queer volunteers I interviewed. A 38-year-old white, bisexual woman shared that it seemed difficult for many queer volunteers to identify as a citizen scientist, which prevents them from developing a collective identity and an associated sense of community with the other volunteers in their group. She attributed this to the rather impersonal focus on data collection and analysis rather than promoting and fostering social interactions based on her experiences with CCS projects and programmes. She shared:

> I don't even think that all of them [queer volunteers] would necessarily call themselves citizen scientists or think of that as a community that they're involved in ... they don't want to claim that as their identity, probably because ... it's so much about collecting data or analyzing data and the social aspect is on the backburner ... we can't find other people as easily to get to connect with and have positive interactions.

Some queer volunteers also shared that the CCS projects they engaged with did not allow space for them to have deeper interactions that would allow them to develop beyond surface-level relationships with fellow volunteers. It seems that in their experiences the project leaders maintained the interactions among volunteers within the context of the project as opposed to promoting more informal connections. A 72-year-old Native American, Two-spirit/lesbian shared about her limited experience interacting with fellow volunteers involved with the project compared to the leaders. She described:

> You still feel like you're part of something, but you don't get to individually get to know the [other] people – you just get to know the educators and the coordinators, but not the individual people who were there.

Despite some barriers to fostering a sense of community, some of the queer CCS volunteers also reported opportunities for community that have great potential within CCS. There were multiple mentions of how shared interests, values, beliefs, and worldviews among volunteers served as points of commonality that could lend themselves to a foundation of community. Given that most CCS projects focus on a specific scientific topic, it is safe to assume that most if not all volunteers have some level of curiosity about that focal element. When asked about what makes something feel like community, a 38-year-old white/Asian, queer/bisexual woman shared, "I would say mutual interests and areas of concern, and some degree of support". In some cases, the community element of CCS can be a motivator for volunteer engagement for individuals seeking a group of passionate, like-minded people. A 73-year-old white, bisexual woman described:

> The community aspect of it was very important to me ... we had a common purpose, we were all advocating for the beach protection and proper zoning of the beach. We all shared this common interest – the common interest of science.

Opportunities for queering citizen and community science

The notion of queering traditional approaches to science and CCS was explicitly expressed by the LGBTIQA+ volunteers I interviewed. In addition to acknowledging that their queer identities may have influenced their approaches to conducting science methods and activities, they also identified the need for opportunities related to queering scientific norms, including leveraging CCS as a tool to do so. A number of individuals describe CCS as a space that allows for individuals with identities outside of who is a 'traditional' scientist to do science. Some of the personal barriers encountered by LGBTIQA+ volunteers are related to being closeted to some degree as well as not feeling connected to fellow project volunteers. One volunteer I interviewed, the same 36-year-old white, queer/asexual/demisexual and agender/non-binary individual mentioned previously, talked about the challenges of finding and forming community with others that they may not share commonalities with beyond their general mutual interest in the focus or topic of their CCS projects. They mentioned that specific spaces that allow them to connect with other queer individuals help them feel a sense of community and empower them to be their full self with others like them, which they have not been able to do or feel with their fellow CCS volunteers. They shared:

> I tend to be pretty closeted about lots of aspects of who I am ... as I've gone longer and longer without really finding other [queer] people like me, without specifically seeking them out, it really does make me feel like an other, like I'm alienated ... when I do have a community, it's great! I'm in a discord server and it's specifically for queer people who are interested in STEM ... it's pretty much the only place I've really felt like I have a community. [Other times] when I've specifically tried to find other queer people I haven't been able to find them – I don't know where they are, I don't know how to connect with other people like me.

Some individuals also noted barriers to fostering a sense of community among CCS volunteer groups by often feeling like outsiders or 'othered' in relation to fellow volunteers based on their queer identities. A 72-year-old Native American, Two-spirit/lesbian who was also previously mentioned revealed one poignant example of this. She described her experiences of participating in a CCS project based in her small coastal community where she was one of the very few non-cishet residents and the only non-cishet volunteer in the project group. She described this lack of connection to her fellow volunteers and community:

> Well they're all heterosexual here – I don't think there's anybody else [who is queer] in the neighborhood ... I want to make it clear that I'm never left out, it's just that there's nothing to bond with. My experiences are so different from theirs.

She expanded further to talk about not only feeling like an outsider based on her sexual orientation and gender identity, but also based on the fact that she was single and without a partner in a small community and project group that was largely made up of heterosexual couples. She said:

> Everybody talks about the people they love ... [project] dinner conversations are in pairs and nobody left me out, I don't mean that, but it's the same everywhere. People talk about their significant other, whoever that is, and they talk about their kids and they talk about their future growing old together ... it's especially disconnecting because I know I couldn't since I'm not heterosexual. There's kind of a double disconnect because it puts me even one step further out.

By considering the personal experiences of LGBTIQA+ volunteers, the field of CCS has the potential to reduce cis-heteronormative barriers and more authentically respect and retain LGBTIQA+ volunteers in CCS. In addition to providing more opportunities for volunteers to form connections with one another beyond project activities and, deeper than surface level interactions, leaders should take more actions to make their CCS projects more queer friendly by directly connecting with LGBTIQA+ organizations to invite them to participate. Additionally, they can implement the practice of actively sharing pronouns, which can demonstrate allyship as well as respect any current queer volunteers, regardless of whether or not they may be 'out'.

It is vital for CCS opportunities to acknowledge the unique ways that volunteers from non-dominant backgrounds, communities, and identities will engage with both the project itself and with fellow volunteers. CCS designers should also consider if, how, and what extent they offer volunteers opportunities to build relationships with one another in addition to how they are creating a welcoming space. Intentional design with an emphasis on inclusion would likely allow volunteers to feel comfortable sharing aspects of their personal identities, which may contribute to their developing deeper connections with fellow project volunteers and result in the cultivation of a volunteer community.

Reflections

The experience of conducting this series of interviews with such a dedicated and passionate group of CCS volunteers was not only insightful and productive in furthering knowledge on the engagement of LGBTIQA+ folks in science, science communication, and broad STEM engagement, but personally gratifying to be trusted with these stories and bonded with these individuals. My biggest takeaway is that LGBTIQA+ CCS volunteers are among the most dedicated and resilient individuals in science. They are often willing to endure personal feelings of isolation, disconnection, and lack of belonging for the sake of furthering scientific research and contributing to science communication efforts that will benefit general public audiences. Despite lacking a sense of community with their project cohorts, queer CCS volunteers remain committed to their participation and contributions for the sake of advancing scientific research and broadening the public engagement, communication, and understanding of science.

Based on my conversations with these 14 queer volunteers, my top recommendations for CCS project leaders and designers would be to:

- *Make your projects queer-friendly.* By acknowledging and validating LGBTIQA+ identities and the value that queer perspectives and approaches add to your project goals and outcomes as well as the broader field of science, you are signalling to active and potential LGBTIQA+ volunteers that they are truly welcome.
- *Create multiple avenues for social interactions.* Communication allows for individuals to become better acquainted and identify areas of commonality. The more opportunities there are to interact in different settings and via different mediums, including informally beyond the scope or context of a specific CCS project, the more potential there is for connections to be made and communities to be fostered.
- *Solicit and incorporate volunteer feedback in project design and approaches.* There is much to be learned and leveraged from the participants who engage with CCS. Inviting volunteers, especially those that identify as queer and/or from other marginalized or underrepresented identities and communities, to express their truth and perceptions regarding their experiences as a CCS volunteer can aid in shaping an overarching project that is welcoming, accessible, and inclusive.

Conclusions

As we continue to make efforts to diversify the field of CCS by reducing barriers and creating safer spaces for individuals often excluded or forced out of STEM, we in turn contribute to a richer public communication of science. A greater diversity of individuals engaging in doing science results in more voices sharing about the science they are pursuing, which can ultimately lead to a shift in how science is perceived by the public in addition to the introduction of more innovative scientific practices. By embracing and valuing LGBTIQA+ identities in CCS and other science engagement opportunities, there is great potential to queer scientific approaches and contribute to creating inclusive scientific practices and a broader STEM workforce.

In science, it is important to recognize who is making the decisions, what their methodological choices are, and how these decisions and choices provide opportunities to queer traditional scientific norms. It is also important to question if and how queer voices and approaches are being centred in addition to those of other marginalized groups. It is vital for the field of CCS to become better acquainted with who these queer volunteers are, what their relationships are with science, and what role(s) their identities play in their relationships with science. By learning from the individuals who directly contribute to CCS efforts, the field will be able to continually evolve to include a broader community of citizen and community scientists.

References

Aristeidou, M., Scanlon, E., and Sharples, M. (2015) 'Weather-it missions: a social network analysis perspective of an online citizen inquiry community', in G. Conole, T. Klobučar, C. Rensing, J. Konert, and E. Lavoué (eds) *Design for Teaching and Learning in a Networked World*, Switzerland: Springer, pp 3–16.

Berger, R. (2015) 'Now I see it, now I don't: researcher's position and reflexivity in qualitative research', *Qualitative Research*, 15(2): 219–234.

Bonney, R., Cooper, C.B., Dickinson, J., Kelling, S., Phillips, T., Rosenberg, K.V., and Shirk, J. (2009) 'Citizen science: a developing tool for expanding science knowledge and scientific literacy', *BioScience*, 59(11): 977–984.

Cooper, C.B. and Smith, J.A. (2010) 'Gender patterns in bird-related recreation in the USA and UK', *Ecology and Society*, 15(4): 4.

Davis, L.F., Ramírez-Andreotta, M.D., and Buxner, S.R. (2020) 'Engaging diverse citizen scientists for environmental health: recommendations from participants and promotoras', *Citizen Science: Theory and Practice*, 5(1): 7.

Dawson, E. (2018) 'Reimagining publics and (non) participation: exploring exclusion from science communication through the experiences of low-income, minority ethnic groups', *Public Understanding of Science*, 27(7): 772–786.

Falk, J.H. and Dierking, L.D. (2010) 'The 95 percent solution', *American Scientist*, 98(6): 486–493.

Farland-Smith, D. (2012) 'Personal and social interactions between young girls and scientists: examining critical aspects for identity construction', *Journal of Science Teacher Education*, 23(1), 1–18.

Frost, D.M. and Meyer, I.H. (2012) 'Measuring community connectedness among diverse sexual minority populations', *Journal of Sex Research*, 49(1): 36–49.

Ganzevoort, W., van den Born, R.J.G., Halffman, W., and Turnhout, S. (2017) 'Sharing biodiversity data: citizen scientists' concerns and motivations', *Biodiversity and Conservation*, 26(12): 2821–2837.

Haywood, B.K. (2016) 'Beyond data points and research contributions: the personal meaning and value associated with public participation in scientific research', *International Journal of Science Education, Part B*, 6(3): 239–262.

Jennett, C., Kloetzer, L., Schneider, D., Iacovides, I., Cox, A., Gold, M., et al. (2016) 'Motivations, learning and creativity in online citizen science', *Journal of Science Communication*, 15(03): A05.

King, A., King, D., Banchoff, A., Solomonov, S., Ben Natan, O., Hua, J., et al. and on behalf of the Our Voice Global Citizen Science Research Network. (2020) 'Employing participatory citizen science methods to promote age-friendly environments worldwide', *International Journal of Environmental Research and Public Health*, 17(5): 1541.

Lewenstein, B.V. (2016) 'Can we understand citizen science?', *Journal of Science Communication*, 14(1): E.

NASEM [National Academies of Sciences, Engineering, and Medicine]. (2018) *Learning Through Citizen Science: Enhancing Opportunities by Design*, Washington, DC: The National Academies Press.

Rahm, J. and Ash, D. (2008). 'Learning environments at the margin: case studies of disenfranchised youth doing science in an aquarium and an after-school program', *Learning Environments Research*, 11(1): 49–62.

Shirk, J.L., Ballard, H.L., Wilderman, C.C., Phillips, T., Wiggins, A., Jordan, R., et al. (2012) 'Public participation in scientific research: a framework for deliberate design,' *Ecology and Society*, 17(2): 29.

Soleri, D., Long, J.W., Ramirez-Andreotta, M.D., Eitemiller, R., and Pandya, R. (2016) 'Finding pathways to more equitable and meaningful public-scientist partnerships', *Citizen Science: Theory and Practice*, 1(1): 9.

Sorensen, A.E., Jordan, R.C., LaDeau, S.L., Biehler, D., Wilson, S., Pitas, J-H., and Leisnham, P.T. (2019) 'Reflecting on efforts to design an inclusive citizen science project in West Baltimore', *Citizen Science: Theory and Practice*, 4(1): 13.

Teaching Notes for Part III

Coming out is not an easy, once-off event. LGBTIQA+ people 'come out' continually in each new context they encounter, once they have gauged how safe coming out might be. People do not have to come out and, if they do, they deserve to come out on their own terms and in their own time. What can we – outsiders to this process and decision – do, to make the environment around them welcoming and friendly? Teachers have a lot of power in this respect – they lay down the rules for engagement and have the ability to enforce them. Science communicators have the same power in the design and set-up of research and practice initiatives.

Questions posed by Part III:

- **Equity, diversity, and inclusion (EDI) in science communication:** as a class, reflect on the visible and invisible aspects of EDI practices and policies in science communication. What is and is not present? How might that affect whether people choose to participate in science communication and how they could be empowered to do so?
- **Setting the rules for engagement:** how do you set the rules for interaction and discussion in your science communication context? How do those rules empower people to engage and participate? How might this be improved? What rules might you want to introduce specifically to address the needs of LGBTIQA+ people?
- **Supporting identities outside 'the norm':** science communication can perpetuate notions of what is normal, including perpetuating norms of sexuality (usually sexually active and heterosexual), gender (as binary and cis), and sex characteristics (as a binary division between 'female' or 'male'). It also perpetuates norms in terms of other axes of power such as ethnicity or 'race', ability and disability, neurotypical versus neurodiverse status, and more. In what ways can science communication flip that script? How do we 'queer' what is 'normal' so that marginalized identities are foregrounded as the norm in science communication and STEM?

Activities:

- **Telling stories with empathy:** we can further visibility for LGBTIQA+ people in STEM and science communication by writing or recording profiles of 'out' individuals. However, public storytelling comes with risks for queer folk. Have your students reflect on how best to ethically and empathically engage with people to tell powerful stories. Ask them to reflect on how to enable their interviewees' agency, and be clear on the importance of consent. Then work together to curate a collection of profiles relevant to your institution or field of interest, in ways that maximize ethics, integrity, and interviewee agency.
- **Design an inclusive event:** have your students work in groups to design (and possibly deliver) a science communication event that is actively inclusive of LGBTIQA+ people. Consider: how can events be made inclusive for all parties? How can this inclusive approach be communicated to the intended audience? Should the organizers revise their assumptions about who the audience is?
- **Diverse event formats:** events can easily attract a singular type of audience (extroverted, non-LGBTIQA+, and so on). When brainstorming ideas for your science communication event, have your students interrogate what alternate formats might work better for a wider array of audiences. Include attention to sub-communities within the LGBTIQA+ rainbow. What kinds of diversity are present within the queer umbrella? Who is often ignored or neglected and how might their interests be served? Map out steps to address these issues within your event.
- **Encouraging feedback:** science communication initiatives can be time-poor and cash-strapped. These factors can limit attempts to gain feedback. Yet, without feedback, it is difficult to know what impact has been achieved. When designing and planning your science communication event, ask your students to design a feedback mechanism that encourages volunteers or attendees to reflect on the event's impact. How can that feedback address questions of inclusion and exclusion in addition to other science communication considerations?

PART IV

Queering Institutional Science Communication Agendas

9

Science OUTreach: A Queer Approach to Science Communication Practice

Alice Motion and Lee Wallace

Introduction

In recent years, forms of public science communication that overtly intersect personal and professional identity have become more commonplace. This intersection represents a rich opportunity to connect science to audiences who have been historically excluded and to experiment with different forms of expression. Of course, efforts to do or share science in cultural contexts that break free of a white, heteronormative, and patriarchal establishment are prone to criticism from those who prefer to propagate these traditions. Of these critiques, perhaps the most perplexing and anachronistic are those that position personal or political narratives at odds with the requirement for 'objectivity' in science (Saini, 2020). While objectivity is paramount to the design of reproducible scientific experiments and unbiased interpretation of results, we must recognize that scientists are people shaped by their life experiences. Good research should not require scientists to leave their identity at the laboratory door. This is especially important for people whose identities are not well represented in lab culture.

Mainstream international efforts to include underrepresented people in science first centred on gender equality with the United Nations declaring the first International Day of Women and Girls in STEM in 2015 (United Nations, n.d.). There is much important work to be done to include women in science, but some of these more widely adopted diversity initiatives can leave 'other aspects of identity side-lined' (Moore and Nash, 2021, p 4), can perpetuate the myth of a gender binary, and can ultimately be exclusive. Visibility for other groups of people who have been historically excluded from science is growing, however, and has, in some cases, been assisted by campaigns starting on social media that coincide with broader acknowledgements of these communities.

In common with other minority groups, LGBTIQA+ people and concerns are underrepresented in science and science communication research and practice (Roberson and Orthia, 2021). In recent years, there has been a growth in organizations and networks that support queer-identified people working in STEM and international days of celebration or 'observance'. The inaugural international 'LGBTQ+ STEM Day' was held in July 2018 but is now formally ensconced as an annual event on 18 November, a date symbolic of the anniversary of the US Supreme Court fight against workplace discrimination by gay activist and astronomer, Frank Kameny (Stenhoff, 2020). In 1957, while working as part of the US space programme, Kameny was summoned back from fieldwork, interrogated about his sexuality, and subsequently dismissed and banned from working for the federal government. Although Kameny failed in his attempts to sue his employer and had his subsequent appeal to the Supreme Court denied, he became a lifelong and celebrated gay rights campaigner until his death in 2011, with additional honours bestowed posthumously.

There have undoubtedly always been queer scientists, but discrimination such as that experienced by Kameny has meant many have remained closeted, a trend which persists in professional contexts today. Indeed, a 2013 survey of LGBTQA professionals working in science, technology, engineering or mathematics (STEM) in the United States revealed that 43 per cent of respondents were not out at work (Yoder and Mattheis, 2016). While the reasons that so many queer people in STEM remain closeted are undoubtedly complex, the sheer number affected signals the requirement for greater inclusivity in these professions and events, such as LGBTQ+ STEM Day, that may engender increased visibility and change.

As a queer scientist and science communicator (Motion) and a queer academic in gender and culture studies (Wallace), we are particularly drawn to the ways that community-led initiatives provide opportunities for the amplification of pre-existing queer science communication events, activations, and performances that have emerged in the burgeoning area of queer science communication practice. In this chapter we draw on this emerging area to explore the ways that queer forms of expression have been applied to the communication of science. We consider how queer people can draw on their lived experiences to enhance the way that science is communicated and discuss how we might better combine theory and practice in this space.

Queer expression and science communication
Stories of queer being and science

Storytelling – or narrative – is a central aspect of science communication and queer people have unique contributions to make in this area because we are compelled and sometimes forced to create narratives on our lives.

Queer people have unique contributions to make in this arena because we are so often compelled to narrate our identity to ourselves and others. Indeed, one of the foundational insights of queer theory, and the gay and lesbian studies that preceded it, is that homosexuality – unlike heterosexuality – requires a narrative or an account of its own becoming (Weeks, 1977; Roof, 1996). This also holds true for people who are not cis-gender. Straight cis-children may grow into their sexuality and gender unannounced, but gay, trans, and other queer people must avow their difference, usually by way of coming out. As a result, the coming out story is a mainstay of queer social worlds, a means of understanding oneself and connecting with others who also share the human investment in narrative as a way of making sense of events and stimuli that might otherwise be experienced as random or inexplicable.

An ability to share stories about science or to construct complex information into compelling narrative forms is a key skill for science communicators, so there is seemingly potential in bringing this queer narrative capacity into the sphere of science communication. Scholars have argued that narrative forms of science communication increase non-expert audiences' interest and engagement (Dahlstrom, 2014). Tension and drama build suspense and excitement which capture audience attention, emotion can build empathy and connection, and all stories have the potential to transport audiences to places that can only be reached by imagination; something that can be important when trying to conceptualize abstract scientific concepts.

As Joubert, Davis and Metcalfe write, 'All animals (and many plants) communicate in one way or another, but only humans tell stories' (2019, p 1). Stories can transcend culture and language when they expand beyond the use of a narrative arc to communicate change, but also effect change in their audience through emotional connection. Queer science communication has the capacity to effect this change through forms that interweave science or research with personal narratives. When stripped back to their simplest form, many emergent queer science communication events include accounts of a queer scientist's becoming as part of the event, whether explicitly as part of the narrative or through vignettes of experience or the world lens of the narrator.

One such nascent narrative event was the Sydney iteration of *500 Queer Scientists* (see Practice Spotlight: 500 Queer Scientists), which has been programmed to coincide with the Sydney Gay and Lesbian Mardi Gras for the past three years. The event has been emceed from its inception by the first author (Motion), who has for many years been an out queer scientist *and* a science communicator (Motion, 2021) but had not strayed into overtly queer spaces to share science until relatively recently. The event takes place in the historically and culturally significant Royal Botanic Garden, Sydney, a place that once hosted Charles Darwin and now provides a safe space for queer participants to share personal stories about science in a community atmosphere. As described by Motion and Sauquet, the event is an example of science communication facilitating new community connections. Audience feedback suggests that this event has helped build community among queer attendees who report previous feelings of exclusion from Mardi Gras and other queer festivals, which are often centred around parties, parades, and protest rather than the pursuit of scientific knowledge. Raising the visibility of queer science and queer scientists is an important step to making science more inclusive, but of equal value is the capacity of events such as this to highlight science as an underrepresented part of queer culture.

Coded stories and learning from queer 'jargon'

Another key feature of good science communication is the ability to adapt material for different audiences and context. That is, science communicators must be proficient code-switchers. Again, queer science communicators may have pre-existing personal expertise to draw from when contributing to this domain.

Code-switching has been variously defined but broadly describes an individual's ability to switch between two or more languages or dialects when speaking. More recently, definitions have expanded beyond the linguistic to include an ability to switch expression, behaviour, and appearance. Here, the code-switcher invariably changes the way that they present themselves to optimize the comfort of their audience and to hopefully engender a favourable response. Minority groups, in particular Black people, First Peoples and People of Colour, have long employed code-switching to 'blend in' within historically white spaces. Because sexuality and gender have complicated relations to social visibility, queer people are also often adept at 'code-switching' as they move from queer-friendly spaces to potentially hostile environments.

To take an extreme example, in the decades when homosexuality was a criminal offence in the UK, gay men developed a coded way to speak to each other about sex while avoiding detection by others (Baker, 2002). The resulting dialect, Polari, is made up of cant, rhyming slang, and nonsense phrases that derive from different linguistic forms historically used by Roma communities, sailors, thieves, sex workers, and Mollies (effeminate men who congregated in particular public houses in London in the 17th century) that can be traced back in English and other European languages as far as the 16th century (Norton, 2002). Polari more or less disappeared from use after homosexuality was decriminalized in the late 1960s, although it publicly persisted as a code for sexual innuendo within the humorous BBC radio show, *Round the Horne*, which reached audiences of up to 20 million (British Broadcasting Corporation, 2018). As this trajectory demonstrates, Polari was not just about secrecy and safety, it was also a means to unite people in their differences (Baker, 2020) and ultimately convey those differences to others in a socially approachable form. Other queer vernaculars persist or emerge within LGBTQIA+ cultures and have been explored by scholars of 'Lavender Linguistics', a term coined by William Leak in 1993 (Jones, 2021).

More broadly, code-switching is an adaptive behaviour – the capacity to stand out, or blend in, in order to secure a range of different effects in different contexts – and a resource that is often put to use in queer instances of popular culture directed at broad audiences. For example, queer television hosts Graham Norton and Ellen de Generes have media personae that rely on likeability and trust with both queer and cis-het audiences. So do the *Queer Eye* lifestyle team who serially makeover both cis-het and queer people into happier, authentic, and affirmed versions of themselves. In part the result of a history of homophobia, transphobia, and biphobia from within both cis-straight and queer communities, this queer capacity to engage others with humour and empathy can also be used to assist the translation of the technical language of science into forms that are more accessible to various publics, particularly narrative forms. Science communicators can also learn how, from the behaviours of effective queer hosts, to welcome lay publics into their scientific 'homes' and assist them to feel at ease with the unfamiliar.

Speaking in code has been a form of safe passage for queer people, allowing them to connect with like people undetected by hostile others. In public science communication, however, speaking in code is now regarded as a problem. Researchers, at least within their home disciplines, mostly understand each other well enough but their language is often beyond the comprehension of even the most interested members of the general public. Unlike the historical use of Polari, code (or jargon as it is more often called in science communication) is neither funny nor useful outside its laboratory home (Rakedzon et al, 2017; Shulman et al, 2020). The code-switching ability of queer people (and other minority groups) provides an ideal – albeit unintentional – training ground for science communication as it supports an ability to move from conversations with colleagues that are steeped in specialist language to authentic and accessible conversations with the public. This involves more than the removal of technical and linguistic barriers to understanding, as critics of the one-way deficit model of science communication have rightly pointed out (Nisbet et al, 2009; Simis et al,

2016). Simply sharing science facts or sound bites with an audience assumed to be lacking in science understanding can be ineffectual, patronizing and alienating. Inherent in the deficit model is a problematic hierarchy that maintains scientists in the position of experts and the public as novices, rather than finding a common ground for dialogue between scientists and members of the public who are experts in their own domain.

Performance in queer science communication

It has long been acknowledged that performance can play a useful role in science communication in, for instance, demonstrations of scientific principles or research within schools and museums. Queer performance traditions, however, often deliberately tap into the celebratory and exploratory role of camp excess as a more open-ended communication strategy. While the story-driven event *500 Queer Scientists* event at the Botanic Garden derives its power from personal authenticity and accessible language, other forms of queer science communication, including some spotlighted in this volume, take a different approach. Instead of appealing to transparency and sincerity, they describe tapping into camp, a queer performance tradition that celebrates the inauthentic or the hyper-real, in order to further the aims of science communication. Indeed, science communication through drag (a subset of camp that has historically featured cross-dressing and gender impersonation but has taken on a different profile in the context of the trans rights movement) can be seen as an emerging practice in several national contexts. 'Science can be a drag, but it doesn't have to be,' reads the tagline of a 2022 event by the UK-based Science is a Drag team (Science is a Drag, 2022; see also Practice Spotlight: Science is a Drag). Blending drag, camp, and cabaret modes, Science is a Drag performers break down complex scientific ideas and make them accessible for public audiences in both in-person and streamed delivery formats.

The rise and popularity of camp science speaks to an element of performance that is key to scientific presentation. In science communication settings and elsewhere, scientists have reflected on the symbolism of the laboratory coat as a professional costume worn by those adopting the role of scientists. In this regard, the lab coat can be viewed as a form of drag that can be subverted by secondary citation, as in the camp approach taken by some drag science communication performers including the Science is a Drag example mentioned earlier. The perpetuation of the visual stereotype of the lab coat-wearing scientist in film and television means the public is attuned to the reversal or flipping of the archetype. Drag performances can, for example, either employ costumes that are not usually associated with the stereotype of a scientist or poke fun at the wild-haired, test-tube wielding, cis-gendered male scientist. An example of this is the work of Lana Vuli, the drag queen sci com performer whose alter ego, Òscar Aznar-Alemany, authored a spotlight for this book. Through Lana, Aznar-Alemany crafted a drag burlesque act about seafood safety that was performed live and via hybridized online events. As can be seen in these examples, drag can move beyond the taking down of stereotypes in order to help interrogate the work of naturalistic performance traditions, such as straight-to-camera speech or the documentary voiceover perfected by David Attenborough, and the role they play in credentialing scientific endeavour and, increasingly, generating scientific celebrity.

Criticisms and costs

There are challenges too for the queer forms of science communication highlighted in this chapter. While narrative is a powerful tool for science communication, scientists are often wary of 'story' as a tool for science communication, thinking it disingenuous or disruptive to the 'objectivity' of scientific data transfer. Science communication workshops or training can help to empower scientists to weave research data into compelling narrative structures without compromising the theories or data that they present. Dahlstrom (2014, p 13617) discusses some of the perceived benefits and challenges of narrative forms of science communication. Positives include the form's ability to 'sway beliefs about numerous science topics' and the potential to shift opinions of 'resistant' audiences to important scientific issues but Dahlstrom also suggests that inaccuracies in narrative forms of science communication can be more challenging to counter. In common with all forms of science communication, the key is therefore to ensure that the science shared is accurate and clearly enunciated within an engaging narrative.

Similarly, code-switching itself can come at a personal cost for queer scientists and other people. For queer people, this may mean modifying our behaviour to conform to expectations of cis-heteronormativity, which can take a significant psychological toll. Recall the over 40 per cent of surveyed LGBTIQA+ scientists who were not yet out at work and were therefore keeping a part of their identity hidden. Although code-switching provides a platform for effective science communication, it could be viewed as problematic to celebrate a behaviour that is perhaps regarded by many queer scientists as necessary for their professional safety and wellbeing.

Reflecting on the two case studies explored – '500 Queer Scientists at The Royal Botanic Garden' and drag forms of science communication – we acknowledge that while the first event featured queer researchers speaking to an overwhelmingly queer audience, the format itself is guided by a tradition that could hardly be considered queer. Although LGBTIQA+ scientists could present their research and personal stories in an approachable way that nonetheless challenged established dynamics and assumptions about the relation between science and non-normative sexuality, the overarching framework for this was professional science. This point was underscored by the enabling partnership with the Royal Botanic Garden, which lent reputational capital to the queer endeavour and broader agenda of inclusivity. The event's success is clear against the conventional goals and objectives of science communication and outreach in engaging a new public for conversations around science and also in increasing visibility and community. These positive outcomes – visibility, community building, and the promotion of science – are consistent with the overall aims of science communication outreach while also representing an extension of that agenda into new space.

Drag forms of science communication push the boundaries of traditional science communication through the creation of overtly queer spaces that are highly inclusive of difference. At their best, these safe spaces allow queer science to be shared in ways that do not shy away from past or present discrimination, problematic politics, or historical erasure, even as they celebrate the scientific contributions of underrepresented groups. As discussed by Roberson and Orthia (2021), queer people have historically been positioned as 'objects' of science with same-sex attraction and gender diversity considered developmental deviations that invite a causal narrative and, often, empirical research and intervention. Queer people continue to contest these positionings on many activist and community fronts and there are now many examples of a 'good' science of queerness that explores animal and plant behaviour without normative presumptions or that centre LGBTIQA+ patients as partners in medical research. Events that focus on research in science without a queer-friendly lens can, however, perpetuate harm for LGBTIQA+ people by supporting problematic narratives of biological determinism and also by erasing the many contributions queer people have made to science, both professionally and as participants in studies.

Looking to the future

To ensure that queer science communication prospers, there are a number of elements that should be fostered. Most importantly, to avoid perpetuating further harm to LGBTIQA+ or other historically excluded people, inclusivity is key (Dawson, 2018; Canfield et al, 2020; Judd and McKinnon, 2021). Participants and facilitators should represent a range of different sexual and gender identities, cultural and ethnic backgrounds, disabilities, and ages, as well as context-relevant disciplines. This is often hard to achieve but the need for inclusive curation cannot be underestimated. Invitations to experts and open calls for expressions of interest to broader contributors must attend to these factors or be found failing against standards that would apply in any other scientific context.

In common with all forms of science communication, those designing queer science communication events need to know their audience and the reasons for engagement. Organizers and performers should also seek to move beyond reimaginations of traditional events that build recognition around queer contributions to professional science, towards events and activities that have an integrated queer identity, form, and location. Deriving from this is the capacity to avoid talking down to either participants or the wider audience. In this context we return to the idea that code-switching can be key to managing complex situations with integrity and respect. In the science communication space, like many other educational contexts, the name of the game is no longer the expert delivery of specialized content but the ability to learn and adapt to the knowledge and experience of others who may not speak in the established idiom of our disciplines. This is where better science communication can ultimately lead to better science and the generation of new and collectively held knowledge.

Finally, as has been widely discussed elsewhere within the science communication literature, queer science communication should encourage a spirit of reflexivity that enables better collaborations between researchers and practitioners (Riedlinger et al, 2019). As an up-and-coming area, queer science communication can reset the relationship between theory and practice by encouraging fluid code-switching between scholarship and performance.

References

Baker, P. (2002) *Polari – The Lost Language of Gay Men*, London: Routledge.

Baker, P. (2020) *Fabulosa!: The Story of Polari, Britain's Secret Gay Language*, London: Reaktion Books.

British Broadcasting Corporation. (2018) 'Polari: the code language gay men used to survive', *BBC Culture* [website] 12 April, Available from: https://www.bbc.com/culture/article/20180212-polari-the-code-language-gay-men-used-to-survive [Accessed 11 February 2022].

Canfield, K.N., Menezes, S., Matsuda, S.B., Moore, A., Mosley Austin, A.N., Dewsbury, B.M., et al. (2020) 'Science communication demands a critical approach that centers inclusion, equity, and intersectionality', *Frontiers in Communication*, 5: 2.

Dahlstrom, M.F. (2014) 'Using narratives and storytelling to communicate science with nonexpert audiences', *Proceedings of the National Academy of Sciences*, 111 (Supplement 4): 13614–13620.

Dawson, E. (2018) 'Reimagining publics and (non) participation: exploring exclusion from science communication through the experiences of low-income, minority ethnic groups', *Public Understanding of Science*, 27(7): 772–786.

Jones, L. (2021) 'Queer linguistics and identity: the past decade', *Journal of Language and Sexuality*, 10(1): 13–24.

Joubert, M., Davis, L., and Metcalfe, J. (2019) 'Storytelling: the soul of science communication', *Journal of Science Communication*, 18(05): E.

Judd, K. and McKinnon, M. (2021) 'A systematic map of inclusion, equity and diversity in science communication research: Do we practice what we preach?', *Frontiers in Communication*, 6: 744365.

Moore, R. and Nash, M. (2021) 'Women's experiences of racial microaggressions in STEMM workplaces and the importance of white allyship', *International Journal of Gender, Science and Technology*, 13(1): 3–22.

Motion, A. (2021) 'Pride in science', *Chemistry World Magazine* [online] 11 November, Available from: https://www.chemistryworld.com/opinion/pride-in-science/4014647.article [Accessed 26 March 2022].

Nisbet, M.C. and Scheufele, D.A. (2009) 'What's next for science communication? Promising directions and lingering distractions', *American Journal of Botany*, 96(10): 1767–1778.

Norton, R. (2002) 'Queer Language', *A Critique of Social Constructionism and Postmodern Queer Theory* [blog] Updated 2 July 2011, Available from: http://www.rictornorton.co.uk/social23.htm [Accessed 26 March 2022].

Rakedzon, T., Segev, E., Chapnik, N., Yosef, R., and Baram-Tsabari, A. (2017) 'Automatic jargon identifier for scientists engaging with the public and science communication educators', *PLoS ONE*, 12(8): e0181742.

Riedlinger, M., Metcalfe, J., Baram-Tsabari, A., Entradas, M., Joubert, M., and Massarani, L. (2019) 'Telling stories to build collaboration between science communication scholars and practitioners', *Journal of Science Communication*, 18(05): N01.

Roberson, T. and Orthia, L.A. (2021) 'Queer world-making: a need for integrated intersectionality in science communication', *Journal of Science Communication*, 20(01): C05.

Roof, J. (1996) *Come as You Are: Sexuality and Narrative*, New York: Columbia University Press.

Saini, A. (2020) 'Want to do better science? Admit you're not objective', *Nature*, 579(12 March): 175.

Science is a Drag. (2022) 'Science is a Drag: All You Need Is Love', *Eventbrite* [website] 16 February, Available from: https://www.eventbrite.co.uk/e/science-is-a-drag-all-you-need-is-love-tickets-221315319077# [Accessed 26 March 2022].

Shulman, H.C., Dixon, G.N., Bullock, O.M., and Colón Amil, D. (2020) 'The effects of jargon on processing fluency, self-perceptions, and scientific engagement', *Journal of Language and Social Psychology*, 39(5–6): 579–597.

Simis, M.J., Madden, H., Cacciatore, M.A., and Yeo, S.K. (2016) 'The lure of rationality: why does the deficit model persist in science communication?', *Public Understanding of Science*, 25(4): 400–414.

Stenhoff, M. (2020) 'From astronomy to activism', *Astronomy & Geophysics*, 61(5): 5.31–5.33.

United Nations. (n.d.) 'International Day of Women and Girls in Science, 11 February', *United Nations* [website], Available from: https://www.un.org/en/observances/women-and-girls-in-science-day/ [Accessed 26 March 2022].

Weeks, J. (1977) *Coming Out: Homosexual Politics in Britain from the Nineteenth Century to the Present*, London: Quartet Books.

Yoder, J.B. and Mattheis, A. (2016) 'Queer in STEM: workplace experiences reported in a national survey of LGBTQA individuals in science, technology, engineering, and mathematics careers', *Journal of Homosexuality*, 63(1): 1–27.

Practice Spotlight: Queer Communicators in Environmental, Climate Change, and Sustainability Conversations

Franzisca Weder

The story of climate change, of destruction and loss, is well represented in mass media around natural hazards and new scientific data. In contrast, new concepts of restoration, eco-cultural identities, and sustainability are not picked up widely in public discourses. There is a similarity here with the fact that the perspectives of non-government organization communicators, activists, and queer science communicators are not often heard or visible in the media. This spotlight focuses on queer communicators showing alternative pathways to overcome the binary between well mediatized climate science and – so far – less told stories about a more sustainable future.

Over the past decades, stimulated by digitalization and related changes in the media ecology, new communicator roles have emerged with different degrees of influence on public discourses. These include bloggers, influencers, and storytellers who influence the way conversations and discourses today are shaped and spread, as they validate and/or challenge topics and tailor information in a certain ways.

Science communication itself has changed as well; formats and actors of science communication are diversifying – stimulated in part by new ways of communication and digital media innovations (Bucchi and Trench, 2021). Equally important for this spotlight is that science communication is increasingly seen 'as culture' (Horst and Davies, 2021), which includes the idea of communication as creation of meaning and (common) sense in various media environments (Weder, 2021).

Social conversations and public discourses are currently dominated by global crises around climate change, biodiversity loss, and migration, but also a massive health crisis (COVID-19). At the same time, they are shaped by the advent of scientific and technological solutions to these problems (for example, applications of artificial intelligence, genetic modification, geo-engineering). At this crucial point of time, queer science communicators can potentially bring in a new perspective in the existing conversations about climate change, referring to an understanding of queering as 'instructional, communicative and performative act which challenges heteronormativity' (Roberson and Orthia, 2021, p 2), which has the potential to create a different meaning.

A research project I led entitled 'Convergence of communicator roles' involved conversations with 25 queer science communicators in Germany, Austria, Australia, and Aotearoa New Zealand in 2020 and 2021. The conversations showed that people who experience changes in their life, who are critical and conscious about injustice, disadvantages, and marginalization are more likely to *problematize* dominant communication conventions. That is, they are more likely to create 'cracks' in existing conversations, in existing patterns of meaning, and in existing structures and scientific narratives. A trans activist interviewee stated that "being green" is "naturally connected to being queer" with both identities fundamentally challenging dominant political paradigms. As a trans museum communicator put it, "Queer allyship is active. As queer and trans climate justice advocates, our fight is deeply personal."

Only a few of the interviewees self-identified as 'professional' science communicators. Most are prolific social media users and they narrate themselves as activists, as ambassadors of change specifically in these new, interactive media environments, or as *curators of social conversations about science*; a gay scientist used the phrase "mediator and curator role". They receive support from peer groups within which they reclaim their activist, ambassador, mediator, or curator identity.

Furthermore, queer science communicators differentiated between the work of "complexity reduction" and "translation" of facts for a wider public on the one hand and queer science communication as *conversational problematization* on the other hand. From their perspective, this includes more than communicating *about* science or facts. Instead, it includes communication explained by the interviewees as "stepping on other people's toes" in the words of a lesbian scientist or as "asking questions, even the awkward ones" as a gay journalist framed it.

A gay science communicator described this innovative perspective on science communication as feeling "advocacy for a certain story and creating possibly revolutionary endings" and embarking on a new, more sustainable conversation, including problematization.

Thus, taking a queer communicator perspective, science communication becomes *advocacy communication*, which means evolving agency and consciousness, taking an active role in shaping the public conversation about science through storytelling and showcasing. There is surely much more to be investigated here in terms of connections between having a marginalized identity and taking an advocacy communication stance. Of course this role is not exclusively reserved for queer science communicators. That is one reason why more research and particularly conversations with diverse marginalized groups of communicators is sorely needed.

References

Bucchi, M. and Trench, B. (2021) 'Rethinking science communication as the social conversation around science', *Journal of Science Communication*, 20(3): Y01.

Horst, M. and Davies, S.R. (2021) 'Science communication as culture: a framework for analysis', in M. Bucchi and B. Trench (eds) *Routledge Handbook of Public Communication of Science and Technology*, London: Routledge, pp 182–197.

Roberson, T. and Orthia, L.A. (2021) 'Queer world-making: a need for integrated intersectionality in science communication', *Journal of Science Communication*, 20(1): C05.

Weder, F. (2021) 'Sustainability as master frame of the future? Potency and limits of sustainability as normative framework in corporate, political and NGO communication', in F. Weder, L. Krainer, and M. Karmasin (eds) *The Sustainability Communication Reader: A Reflective Compendium*, Wiesbaden: Springer, pp 103–119.

Practice Spotlight: How LGBTIQA+ Representation in Organization Leadership Impacts Inclusivity and Visibility

Sarah Durcan and Andrea Bandelli

Science Gallery International (SGI) activates and amplifies the third mission of universities: social impact, public engagement, and research influence, strengthening the relationship between society, and forward-thinking academic institutions. We provide the know-how, services, tools, and resources required to catalyze the growth of the Science Gallery Network, the world's only university network dedicated to breaking down the barriers between science and art. We are based in the queer-friendly and socially progressive city of Dublin, Ireland.

Recognizing that science and technology are too often exclusionary or diminishing of LGBTIQA+ people and experience, it is vital for us at SGI, from our position of powerful inclusivity, to model and promote inclusive and LGBTIQA+ visible science communication. We do this so that more diverse people can see themselves and be themselves in the expanding possibilities of art and science, through working with universities to help them shift their approach to education and public engagement. We act in a 'queer way' by challenging traditional establishments (universities), prioritizing certain activities (wellbeing, mutual respect, and so on), and embracing unconventional ways of working that value the individual, not the system.

SGI is not a queer organization by mission – we didn't set out purposefully to hire and retain queer staff. Yet our entire senior management team identify as LGBTIQA+ and we believe that queerness suffuses our approach to everything. Our subjective experience of working in an organization with a high proportion of LGBTIQA+ employees is more than the absence of discrimination or fear: our team members describe it as fully inclusive, empowering and psychologically safe; and an opportunity to bring our full selves to work. We work to create this psychological safety, recognizing the absolute importance to everyone's wellbeing of feeling equal and being accepted, included, and celebrated for all they are. We do this by having honest and challenging conversations, making sure everyone is listened to with respect, and trusting each other.

There is also LGBTIQA+ representation on our board and in senior management across our network locations. We realize that a critical mass of LGBTIQA+ representation at a senior level is vital to ensure that queer perspectives are not 'the voice of the queer person' but have a significant weight across the whole team. Our leadership team values and encourages our staff to be visible activists on their own time (volunteering with other queer and feminist non-profits and movements) and we share and talk at work about these experiences. In this way, we become role models for each other, and strengthen an inclusive organizational culture that is quickly absorbed by new staff. We especially consider how the tools and approaches of our everyday activism may impact and continue to shape science communication across our network.

To us, transdisciplinary work challenges heteronormativity. Queerness is well aligned with the breaking down of silos, wiring new networks, developing new forms of thinking, and education and creativity, and bringing innovation to cultural and science communication formats. Our queer perspective is constantly translating between worlds/communities where we are safe and included, and worlds that were not built to include us; between the centre and the periphery, the minority and the majority; and can powerfully intersect with feminism, anti-racism, and decolonization. And yet, we still face challenges given the make-up of our team. Working in an international context, and seeking to expand the Science Gallery Network, we have to consider the safety of our team members if we were to work in countries where LGBTIQA+ people are legally threatened and discriminated against by the state.

This balance of formal and informal ways of enacting our queer identities manifests itself in the initiatives we spearhead and support. We do not fear or hesitate to support LGBTIQA+ issues, because we can speak from personal experience, and put the authority of our leadership behind them. One example: in our meeting room while making signs for the March for Science 2017, we started a conversation about what we could do to support more LGBTIQA+ visibility. We then reached out to our network of LGBTIQA+ folk working in science communication, and together we established the LGBTQ+ STEM Day (Bandelli, 2018) that is now a worldwide initiative. We were active in its strategic direction and identifying activities and publicity.

We recommend that if you care about active LGBTIQA+ inclusivity in your STEAM programming and communications, then you need to ensure that there is significant LGBTIQA+ representation in your top management and governance structures. While this is our subjective experience, it would be interesting to have objective comparative studies with organizations where LGBTIQA+ folk don't feel included or safe. This could lead to more concrete recommendations on how to achieve inclusivity across many more science communication organizations.

References

Bandelli, A. (2018) 'The first International Day of LGBT in STEM', *Ecsite Spokes Magazine*, 41(May): online, Available from: https://www.ecsite.eu/activities-and-services/news-and-publications/digital-spokes/issue-41# [Accessed 26 March 2022].

Practice Spotlight: Outer Edge: Queer(y)ing STEM Collections – A Community Workshop

Eleanor S. Armstrong and Sophie Gerber

On 5–6 March 2020, the *Outer Edge: Queer(y)ing STEM Collections* workshop was held at the Technisches Museum Wien, Austria (TMW). The workshop was funded under the museum's 'Focus Gender' programme led by Sophie and co-organized by Eleanor. In this spotlight we describe the workshop as a best practice example that queered the relationship between science communication theory and its practice in museums. All quotes in this text come from the graphic recording and anonymous participant feedback.[1] Queering the formal 'conclusion', we have left a series of prompts after each reflection inspired by our work, through the text to support you, the reader, in adapting this model to your queer science communication community building. The questions are purposefully unanswered, with the aim of directing your work instead of giving answers.

To support queering science museum practice – curation, engagement, and learning – we built a community of practitioners where ideas and practices were exchanged and developed collectively. One participant reflected it was 'the first time having the chance to present it in a conference-type set-up'.

What community need is your workshop/event responding to?

We centred a queer methodology in the workshop. For example, we used live visual recording in a zine to capture thinking-in-action during the workshop for sharing beyond our participants. Some of the text sections in the zine were written by us, some by the zine artist, and some are reflections or quotes from participants. We used this instead of video recording (which can be dangerous to expose closeted queer participants, or those working under homophobic regimes) or a written report (which privileges a single voice).

In what ways might you be able to share or develop multi-vocal recordings of your workshop that are safe for your participants?

Our physical-digital hybrid brought together international scholars and practitioners. For example, Rajni Gupta analysed gendered dimensions of the displays in Gujarat Science City, India; Martha Clewlow showed how heteronormative tagging can make queer histories (in)visible in UK science museum databases; and TMW colleague Ana Daldon presented her work developing the exhibition *Who owns Pink?*. One respondent indicated that the range of sessions was 'helpful to understand the different dimensions' for queer practice in STEM collections.

How can you integrate contributions from external and internal participants in your workshop?

Further, rather than participants sharing only practice-past, we supported practice-in-action sessions. Groups of participants (including TMW staff) responded creatively to existing displays at the TMW. This facilitated collective, collaborative learning that dismantled perceived barriers between professionals from museums, universities, activist groups, and students; facilitating community building.

How could you support community building where participants from diverse professions are equally/equitably valued?

We specifically attended to what 'gendered' and 'heteronormative' STEM looked like in the collections. In reflections on language participants asked: 'When we talk about gender why do we almost always talk about women?' When thinking about the terms queer and gender, they asserted we should 'never essentialize these terms', but see them as 'constantly evolving, fractal, emergent.'

How do you support language use (especially in multilingual groups) in the workshop?

The workshop highlighted how gender and sexuality are part of larger systems of STEM. Participants reflected that the TMW's Measurement and

Surveillance displays were framed by a 'narrative of progress' that paralleled some public narratives of queer acceptance as 'modern' and therefore 'progressive'. This allowed us to think about how to place queer stories in historic STEM collections without relying on rhetorics of modernity. STEM collections' existing racialized biases were also brought out: a participant reflected 'we want to see not just white people' as part of queer curation.

What collections that are not normally seen as 'queer' could you use for discussions in your workshop?

We bridged perceived divides in research and practice successfully with participants observing that the workshop 'combine[d] academic and non-academic work'. Our queer methodology sought community direction for future iterations and priorities. Participants specifically noted that 'disability', marginalized queer communities (for example, 'rural communities', 'asylum seekers', and 'prisoners'), and 'intersectionality' were underrepresented and a focus on developing skills in 'ethical values' would be pertinent. These will act as points of departure for future workshops.

How can your workshop plan for future iterations? Who is guiding their content?

Note

[1] The graphic recording zine 'Queering STEM Collections: 1st Vienna Workshop' is available from: https://www.technischesmuseum.at/jart/prj3/tmw/data/db/dokument/Outer-Edge-final_2020-12-10_1112505.pdf.

10

The Possibilities of Queer in Science Communication Teaching and Pedagogies

Simon J. Lock and Eleanor S. Armstrong

Pedagogy in science communication can be wide ranging, spanning formal education, informal education, everyday science learning, and professional development contexts. This chapter considers how we might apply the concept of queer pedagogy to these multiple processes and settings. To paraphrase Luhmann (1998), what would a queer pedagogy of science communication (QSCP) look like? What would be its ambitions, and where would it take place? Is QSCP something just for queer science communicators and students? Or would the practice also include queering the science communication curriculum? Or is it about queer learning, teaching, and practices within the broad umbrella of science communication? In this chapter we approach all these questions and explore what queer theory might bring to science communication and the possibilities of teaching it queerly.

Radical pedagogies (hooks, 2003; Freire, 2018; Giroux, 2020) and queer theories have served as forms of critical or subversive interventions in both oppressive classroom relations and social architectures of heteronormative sexualities and genders for a long time now. Pedagogy is often understood as referring to the 'how-to' of teaching. Yet more recently, as Luhmann notes, 'flagged by signifiers such as *feminist, radical,* and *anti-racist,*' pedagogy is now 'highly critical of mainstream education and of its tendency to reproduce racial, gendered, and class-based power relations in its institutions, ideologies, and practices' (Luhmann, 1998 p 125, italics in the original). Common to these radical pedagogical approaches is the desire to intervene in the reproduction of power dynamics and to create more inclusive learning environments through the transformation of curricula and the structures of social interactions within classrooms. In using queer here we draw on the distinctions made by Morris (1998) between the use of queer as a subject position and queer as a politic, which do not always necessarily overlap. While queer theory emerged from the study of, and viewpoints of, people who are considered outsiders in terms of sexual and gender identities, it has also been used to interrogate all claims of 'normalcy' and the processes by which these are defined and policed (Greene, 1996; Morris, 1998; Shlasko, 2005). If, as Britzman (1995) suggests, queer pedagogy's goal is the radical practice of deconstructing normalcy, then as Luhmann (1998) argues, 'it is obviously not confined to teaching *as, for,* or *about* queer subject(s)' (p 129). A queer pedagogical agenda needs to be extended to encompass the refusal of *any* normalization, be it racist, ableist, misogynistic, or any other forms of oppression. Therefore, through an interrogation of the often implicit assumptions and normalcy of heteronormativity, queer theory 'offers a critique of reigning ideologies of subjectivity, power, and meaning' (Greene, 1996, p 326), all things that we argue here are at the heart of science communication but also very much at stake in all communicative encounters.

In planning this chapter, we were alive to Luhmann's (1998) concern as to whether a queer pedagogy could resist the desire for authority and stable knowledge. That is: could it resist disseminating new knowledge and new forms of subjection from the position of authority? If the goal of queer pedagogy is to destabilize norms and normalcy we felt that attempting to write *the* chapter on this topic would not be in the spirit of either set of theories and approaches. Thus we have approached this chapter in the disruptive spirit of both by posing a series of questions to ourselves and our readers, bringing our own sets of knowledges to bear on them, and fully accepting that these will be limited by our own queer positionalities. Our aim here then, is to model an example of what a QSCP might look like, how it might be experienced and practiced, and to equip our readers with a set of ideas that they can take into their own classrooms, workplaces, and practices.

What follows are some prompts, each followed by responses and reflections from the authors. As such, we felt it important to both be clear with ourselves and our readers what our own positionalities are (at the time of writing). SJL uses he/they pronouns and is a queer white British academic in their 40s, they are genderqueer and polyamorous. ESA uses she/her pronouns and is a queer, dyslexic, white British academic (living in Sweden) in her 20s.

Inclusion of queer people, stories and examples

As Roberson and Orthia (2021) have previously noted there exists an underrepresentation of queer people within science communication both within the examples used, as subjects within the science that is being communicated, and also within those people teaching and practicing science communication.

There are perhaps three axes – reformulating here the prompts in Winchester (2012) for the context of scientific practice – that we might think along in pedagogical practices to introduce LGBTIQA+/queer issues: who are we talking about? Which parts of science are we talking about? And how do we discuss topics that matter to the community? There are a range of organizations around the world that represent queer scientists (for example, Pride in STEM, 500 Queer Scientists, OutSTEM, Queer Scientists PH) providing useful points of departure for science communicators to independently broaden the 'who' within their discussion of the hegemonic canon of science. Bringing QSCP into the 'who' we represent in science, we can also challenge the norms of who constitutes a scientist – bringing in narratives of non-traditional scientists such as people in manufacturing, laboratory or technical works, or practicing outside the strictures of formal 'science' (for instance, Liboiron, 2021). To challenge ideas about 'what' is being considered; instructive books on queering science studies (for example, Cipolla et al, 2017) can act as points of departure for thinking about which queer topics of science can be communicated, and how they are framed within the discourses created through science communication.

QSCP also encourages us to draw attention to issues that are of concern to queer communities (a selection of which are included in this text) – particularly in the communication of medical sciences, and biological sciences where cis-heteronormative bodies and actions are normalized in multiple ways. With better, queerer, representation comes alternative knowledges, with learning about queer scientists and queer publics hopefully comes the realization of their valuable place within science, and we can move towards less discrimination within STEM (Dyer et al, 2019; Cech and Waidzunas, 2021) but also within wider social values. We might also see this approach as a method for addressing other pedagogical concerns around creating an inclusive classroom that supports queer folk.

However, the question is predicated on understanding that there are 'correct' 'places' where we should do this inclusion work, and that assimilatory politics is a desirable outcome. In the first case, we can challenge these 'places' where this inclusion happens. Such categories might include critiquing who the 'public' are – rejecting the single, monolithic category and instead closely attending to queer groups of people who constitute less obvious 'publics'. QSCP also develops our thoughts about who counts as a legitimate voice of 'science' to communicators – what heteronormative (and/or homonormative), racialized, or ablist assumptions might be at play when selecting individuals to represent science or scientists? Using QSCP directs our attention to what we constitute as 'science' – how do the selections of topics in science to communicate confirm the positions of science as 'objective', 'universal'? Thus, QSCP illuminates possibilities for learners to think beyond science, and science communication as being something untied to representation.

Additionally, an assimilatory politics of QSCP is limited. Grounded in a set of assumptions common to older forms of LGBTIQA+ politics, assimilatory work assumes that homophobia is little more than a problem of representation, an effect of non-existent or distorted images of lesbians and gays. The job of representation, therefore, is to counter these longstanding representations of lesbians and gays solely as sick, sexually perverted, unhappy, and antisocial with positive representations of LGBTIQA+ life. This goal of inclusion within the existing structures and binary-gendered heteronorms risks erasing the diverse trans and non-binary lives and queer identities that dissemble the gender binary, as a result of following a homonormative or assimilationist approach to queer politics (Duggan, 2002). To show queer people as what Ahmed (2010) terms 'happy objects' – oriented only toward heteronormative lives, knowledges, and things that make people 'normal' and 'happy' – is a limiting approach. Should we, for example, be happy because we have queer science communicators but ignore their promotion of science research sponsored by the military-industrial state, which underpins violence against marginalized groups worldwide? If we agree that science communication is politics then a QSCP must attend to this dimension as well.

'Queer as politics' approach

As Shlasko (2005) argues, a queer theoretical lens can provide a framework through which we can interrogate not only the treatment of queers but also queerness of identity and normalcy. So that means queering science, queering publics, and queering science communication, and queering the underpinning assumptions about what it is for. A critical understanding of the public sphere – the context for most science communication – and its entrenched power dynamics and oppressions along the lines of gender (Fraser, 1990), race (Dawson, 2019), and sexuality (Warner, 1993) must be centred. These dynamics structure and effect those with marginalized identities and their ability to speak, listen, participate, and be represented in the context of science communication. We need to talk less about 'hard to reach groups' or publics that are 'difficult to engage', and start talking about those groups and identities that are often medicalized via science and problematized by science communicators.

It is not a coincidence that these groups are generally the most marginalized groups within society, which again speaks to structural inequalities not abject publics. The onus is on science communicators to do the work and not perpetuate harmful and reductive stereotypes of marginalized publics as having the wrong attitudes or orientations towards science. We can look at the work Prepster (2020), a PrEP[1] activist group in the UK, have done to recognize the needs of queer, Black, and trans audiences and tailor their outreach in ways that are meaningful to those specific groups rather than blaming them for not responding to existing mainstream health promotions. We might also recognize grassroots science communication in zines on health conditions (see Vigour and Cook, 2018) which centre the lived experience of those living with them, created and produced by people who are experts in their own bodies and experiences. Such items are often able to identify gaps or absences in mainstream advice and support and create networks of care that sit outside of mainstream health interventions.

Science, as we communicate it now, has historically been a large part of the production of norms of the body more generally as well as producing heteronormativity and binarized gendering. Science communication in turn supports and mobilizes this scientific authority. For example, the construction of demonstrations (by and to gender-segregated audiences) of scientists of the Enlightenment period at institutions such as the Royal Society and Royal Institution, which were later exported around the world, have been framed by the power of colonial imperialism as being *the* legitimate method of presenting science in public. This bleeds into the contemporary understandings of what 'correct' science communication looks like: valuing demonstrations and one-way presentation, for example, over artistic engagement or dialogical and collaborative practices. QSCP can challenge the foundational certainty of how science and science communication should be done, to prioritize non-traditional modes of engagement that are developed by, with, and for diverse audiences (Simon, 2016).

QSCP in practice

We have used these ideas in our science communication practices. For example we have helped learners ask questions not just about the representation within contemporary scientific studies, but the premises that underpin such studies too. In the London Science Museum, for example, curators narrativize the original double helix model of DNA as being a way to understand our identities – knowledge of your DNA *is* knowledge of your identity. In our QSCP we looked at research studies that aim to 'discover' 'gay' genes by correlating genetic testing with survey responses from participants recognizing that such studies are never value neutral. We encouraged learners to ask questions such as: who feels comfortable about going to a laboratory and declaring their sexuality to researchers? How was evidence of one's sexuality defined? How might scientific data be shaped by queer subjects performing notions of 'good' or 'normative' homosexuality to researchers? Deeper still; why would knowing if there was a biological basis of 'gayness' be a good thing? Is 'gayness' the same as 'queerness', 'homosexuality', or 'bisexuality'? What is (or could be) the intended use of this knowledge? Applying a QSCP helps unpack the norms we perpetuate around what knowledge is legitimate science.

Underpinning many of these studies is an appeal to the idea that the cultural standard for knowledge is Science (Haraway, 1988); and that understanding any phenomena 'scientifically' makes the knowledge automatically more robust, more objective, and thus 'better'. In news media science communication on studies of sexuality and arousal, for example, where there is poorer agreement between women's genital measurements and participants' self-reported arousal than men, women are framed as deviant: 'women defy categorization' (Er-Chua, 2009), there is a 'discord, in women, between the body and the mind' such that 'with the women, especially the straight women, mind and genitals seemed scarcely to belong to the same person' (Bergner, 2009). A QSCP 'queer as politics' approach instructs us to unpack how science communication develops and maintains cultural narratives of women's sexual problems. While theorists such as Spurgas (2020) have argued that such scientific studies *create* women's sexual desire as a problem to be solved in science, we must be astute to how science communication perpetuates these into the public sphere.

A QSCP approach that emphasizes queer as politics needs to be alive to creating its own norms around queerness too. Fungi, for example, well known to have an incredibly diverse range of sexes, are often used in science communication practices to demonstrate that using the 'lens of life's diversity, we should not be surprised that we humans are variable too' (Rokas, 2018). Thus, in addition to fungi's queerness for investigating the tensions between social norms and scientific expectations (Kaishian and Djoulakian, 2020), we can use QSCP to unpack how these tensions are mediated to publics in science communication making queerness acceptable because of its occurrence in the natural world. However, critique of acceptability as a result of 'naturalness' in the animal and plant world is vital, as without this dimension of a QSCP, science communication overlooks the cultural constructedness of the category of 'natural' in the first instance and its implications (Terry, 1997).

To queer science communication itself then is to be attentive to its role in constructing and perpetuating dominant social norms around gender, race, sexuality, and ability. As we teach in our science communication module for undergraduate students, from day one, science communication *is* political, it *is* power. To pretend otherwise is to ignore science's role in both constructing our ideas about gender, race, sexuality, ability under western capitalist-colonialist societies but also historically justifying these norms directly and indirectly. So let us question what norms we are upholding through our actions as science communicators. Are we faithful cheerleaders for science or can we be critics? We might see this through the lens of queer politics: are we here to make science more palatable for general consumption, to sell science (an assimilationist approach), or are we here disrupt the status quo? To challenge the heteronormativity of science? To disassemble the role science has played in constructing harmful hierarchies of racial difference? To open up science to wider sets of knowledges? QSCP encourages the learner to reflect on the aims and goals of science communication, rather than simply just the content and methods.

Innovating theoretical constructs through QSCP

To approach science communication queerly means to be attentive to how, and why, oppositional binaries are constructed. Historically (particularly under a deficit model approach), any teaching of science communication is structured around a series of binaries. At the core of a lot of science communication work is not just a pedagogy of knowledge transfer, but more broadly the removal of ignorance. Thus key binaries that have underpinned almost all science communication are expert/lay, scientist/non-scientist, knowledgeable/ignorant. Many theorists (for example, Simone de Beauvoir and Edward Said) have argued that construction of the 'one' creates the need for the oppositional or abject 'Other'. Much like the need for heterosexuality to need its abject 'Other' in homosexuality, science has through the active promotion of communication, rhetorically constructed: lay-people, pseudo-science, other forms of knowledge as 'Others' in the pursuit of its own epistemic authority, status, or the seeking of resources (Gieryn, 1983; Lock, 2008).

QSCP asks us to resist these formal binarizations along with other binarizations that are implicit within them or that are underpinned by them and recognize their historical and contemporary fluidity when setting up science communication educative practices. We also can address how we select the binaries that *should* be tackled within science communication (for example lay/scientist), and which ones are outside of this remit (homo/hetero), to avoid positioning queer and science as oppositional. QSCP might instead invite reflection on the role science communication plays in constructing such binaries and the work this is doing in terms of who has power and who might be being othered or oppressed in the processes and activities we teach. Who is being excluded by such binaries? Who is made abject in relation to science?

Developing and implementing QSCP

We should first return to queer pedagogy *beyond* science communication and its call to develop critical learners. Criticality can come in many frames, but importantly we reiterate that queer pedagogy encourages the refusal of any normalization, particularly racist, ableist, misogynistic, and other oppressive regimes of knowledge and categorization in the pedagogical space. An intersectional commitment to QSCP is lost without this. It also aims to expose and undo the structural oppressions inherent in scientific work, it aims to open the door and hold space for other ways of knowing. Thus, a (critical) QSCP must embrace these spaces for ignorance as an opportunity to engage and learn, and not simply correct and blame.

As such, instead of instructional 'tools', we provide here a series of questions to prompt and guide our readers as they go forth and queer their pedagogical spaces in new and exciting ways:

- Who is in your communication project?
- What stories are we telling about science in both the past and the present?
- Which scientists are we valorizing? Are they the cis-white-heterosexual men of the Global North?
- What kinds of science are we missing? Why?
- How can we extend a QSCP beyond the obviously queer locations and topics in science?
- How can queer actions disrupt norms, categories, and practices that are bound up in social structures and reproduced through communication practices?
- What do norms of science communication practices tell us about our motivations for doing science communication?
- How might science itself be implicated in the binaries of hetero/homo, scientist/public, expert/lay?
- How can we elevate narratives from the Global South, indigenous knowledges, as well as knowledges and science in the West that fell outside the science/non-science binary?
- What larger social norms does (your) science communication speak to? How is it unpacking, critically interrogating, or disrupting the reproduction of these norms when you are teaching science communication?
- How can we challenge the mobilization of norms and binaries through science communication?
- What does science (and by extension, science communication) gain from continuing to perpetuate binaries?
- Why are we teaching science communication at all?
- What is a science communication pedagogical process trying to achieve or produce?

We ask our readers to embrace the idea that science communication is never simply about the transference or representation of information but is always fundamentally involved in politics, cultural reproduction, and power. A QSCP is not simply about adding in a smattering of queer scientists to the examples we use, it is about being aware of why they were not there in the first place. It is aware of the power of scientific narratives to shape and constrain our lives under a dominant white cis-heteronormative patriarchy – it is disruptive, decentring, liberatory, and non-binary.

Note
[1] PrEP (pre-exposure prophylaxis) is medicine people at risk for HIV take to prevent getting HIV.

References

Ahmed, S. (2010) 'Happy Objects', in M. Gregg and G.J. Seigworth (eds) *The Affect Theory Reader*, Durham: Duke University Press, pp 29–51.

Bergner, D. (2009) 'What do women want?', *New York Times Magazine*, 22 January, pp 1–12.

Britzman, D. (1995) 'Is there a queer pedagogy? Or, stop reading straight!', *Educational Theory*, 45(2): 151–165.

Cech, E.A and Waidzunas, T.J. (2021) 'Systemic inequalities for LGBTQ professionals in STEM', *Science Advances*, 7(3): eabe0933.

Cipolla, C., Gupta, K., Rubin, D.A., and Willey, A. (eds) (2017) *Queer Feminist Science Studies: A Reader*, Seattle: University of Washington Press.

Dawson, E. (2019) *Equity, Exclusion and Everyday Science Learning: The Experiences of Minoritised Groups*, London: Routledge.

Duggan, L. (2002) 'The new homonormativity: the sexual politics of neoliberalism', in R. Castronovo and D. Nelson (eds) *Materializing Democracy*, Durham: Duke University Press, pp 175–194.

Dyer, J., Townsend, A., Kanani, S., Matthews, P., Palermo, A., Farley, S., and Thorley, C. (2019) *Exploring the Workplace for LGBT+ Physical Scientists*, London: Institute of Physics, Royal Astronomical Society and Royal Society of Chemistry.

Er-Chua, G. (2009) 'Arousing questions about female sexuality', *The Queens Journal* [online] 9 February, Available from: https://www.queensjournal.ca/story/2009-02-09/features/arousing-questions-about-female-sexuality/ [Accessed 2 August 2021].

Fraser, N. (1990) 'Rethinking the public sphere: a contribution to the critique of actually existing democracy', *Social Text*, 25/26: 56–80.

Freire, P. (2018) *Pedagogy of the Oppressed*, New York: Bloomsbury Publishing.

Gieryn, T.F. (1983) 'Boundary-work and the demarcation of science from non-science: strains and interests in professional ideologies of scientists', *American Sociological Review*, 48(6): 781–795.

Giroux, H.A. (2020) *On Critical Pedagogy*, New York: Bloomsbury Publishing.

Greene, F.L. (1996) 'Introducing queer theory into the undergraduate classroom: abstractions and practical applications', *English Education*, 28(4): 325–339.

Haraway, D. (1988) 'Situated knowledges: the science question in feminism and the privilege of partial perspective', *Feminist Studies*, 14(3): 575–599.

hooks, b. (2003) *Teaching Community: A Pedagogy of Hope*, New York: Routledge.

Kaishian, P. and Djoulakian, H. (2020) 'The science underground: mycology as a queer discipline', *Catalyst: Feminism, Theory, Technoscience*, 6(2): 1–26.

Liboiron, M. (2021) *Pollution is Colonialism*, Durham: Duke University Press.

Lock, S.J. (2008) *Lost in Translation: Discourses, Boundaries and Legitimacy in the Public Understanding of Science in the UK*, PhD Thesis, University of London, University College London.

Luhmann, S. (1998) 'Queering/querying pedagogy? Or, pedagogy is a pretty queer thing', in W.F. Pinar (ed) *Queer Theory in Education*, Mahwah: Erlbaum, pp 141–156.

Morris, M. (1998) 'Unresting the curriculum: queer projects, queer imaginings', in W.F. Pinar (ed) *Queer Theory in Education*, Mahwah: Erlbaum, pp 275–286.

Prepster. (2020) 'The requisite project: for HIV and sexual health, non-governmental and community-based organisations', *Prepster* [website], Available from: https://prepster.info/wp-content/uploads/2020/12/The-Requisite-Project-Campaign-Briefing.pdf [Accessed 5 August 2021].

Roberson, T. and Orthia, L.A. (2021) 'Queer world-making: a need for integrated intersectionality in science communication', *Journal of Science Communication*, 20(1): C05.

Rokas, A. (2018) 'Where sexes come by the thousands', *The Conversation* [online] 30 October, Available from: https://theconversation.com/where-sexes-come-by-the-thousands-105554 [Accessed 2 August 2021].

Shlasko, G.D. (2005) 'Queer (v.) pedagogy', *Equity & Excellence in Education*, 38(2): 123–134.

Simon, N. (2016) *The Art of Relevance*, Santa Cruz: Museum 2.0.

Spurgas, A.K. (2020) *Diagnosing Desire: Biopolitics and Femininity Into the Twenty-first Century*, Columbus: The Ohio State University Press.

Terry, J. (1997) 'The seductive power of science in the making of deviant subjectivity', in V.A. Rosario (ed) *Science and Homosexualities*, New York: Routledge, pp 278–302.

Vigour, L. and Cook, N. (2018) 'Six personal health zines that might change your life', *Wellcome Collection* [website] 4 October, Available from: https://wellcomecollection.org/articles/WsT4Ex8AAHruGfWz [Accessed 5 August 2021].

Warner, M. (ed) (1993) *Fear of a Queer Planet: Queer Politics and Social Theory*, Minneapolis: University of Minnesota Press.

Winchester, O. (2012) 'A book with its pages always open?' in R. Sandell and E. Nightingale (eds) *Museums, Equality and Social Justice*, Oxon: Routledge, pp 142–155.

11

Queering Science Communication Theory Beyond Deficit and Dialogue Binaries

Lindy A. Orthia and V de Kauwe

> Content warning: this chapter deploys sexual metaphors and discusses Francis Bacon's use of a rape metaphor to promote science in the 17th century.

What might queer perspectives offer science communication theory? In this chapter, inspired by queer theory's metaphorical potential, we challenge the science/public binary that underlies most of the science communication models at the centre of science communication theory. We focus on models because they constitute the most well developed area of science communication theory (Trench and Bucchi, 2010). We argue that queering science communication models offers liberatory possibilities for science communication by undermining the hegemonic authority of Western institutionalized science and by granting epistemic legitimacy to marginalized people, including LGBTIQA+ people.

In the spirit of queer pluralism, we take two simultaneous paths to arrive at that point. The left option takes a straighter academic path and examines two queer case studies that illustrate key problems with science communication models. The right option takes a queerer path, using an extended metaphor to explore the gendered and sexual dimensions of the relationship between science and science communication to show what is in desperate need of change. Choose whichever you prefer to read first, then read the other if you like. The paths come together at the chapter's end for conclusions applicable to both.

The science/public binary in science communication models

Many typologies of science communication models have been proposed over the years, but a consensus typology based on a review of these proposals (Metcalfe, 2019) distinguishes between three:

- under the *deficit*, model science communicators treat non-scientists as having a knowledge or attitudinal deficit, so one-way communication methods predominate;
- the *dialogue* model prioritizes two-way communication because science communicators recognize non-scientists have views about science and sometimes useful knowledge to share;
- the *participatory* model sees scientists and others as equal participants in knowledge creation and decision making.

Despite their different political inflections, these models are all premised on a binary division between science and its others, where these 'others' are most frequently characterized as various 'publics'. The deficit and dialogue models self-evidently exemplify that binary. In contrast, the participatory model to an extent challenges it by opening communication and knowledge co-creation to many forms of expertise (Metcalfe, 2019). However, the binary is retained even here. For example, Metcalfe (2019, pp 385–386)

The tale of queered sci com

For those already well versed in science communication models, jargon, and norms, let us retell the tale of contemporary science communication in the form of a playful metaphor.

When apt, metaphors grant us clarity and a new language with which to dissect a problem. Science communication theorists may loftily pontificate about ideals for public engagement with science, but when it comes down to it, there is much to be learned by applying lessons from personal relationships to the socio-political domain. It clears the weasel waffle so we can see both the wood and the trees.

A colleague recently emailed our research network about an upcoming seminar at which she promised to discuss 'the horny old deficit model'. After considerable e-laughter, she admitted she perhaps meant 'hoary' or 'thorny'.

Yet there it was.

The idea of a horny old deficit model is, we're sure you'll agree, disgusting. But what does it look like if we personify it? Perhaps for you (as for us) the phrase evokes a stale pale male academic who can't stop perving at younger colleagues. He bails them up in tight corners, a little too closely, to share his views on all manner of things they didn't want or need to know about.

describes the participatory model as signalling 'a shift in power ... from the scientists to the public', with some of its aims being to 'Participate with various public[s] in policy-making, and integrate their views', to 'Participate culturally with interests other than science', to 'Shape the scientific research agenda', and to 'Critically reflect on science and its institutions'. Science thus retains its distinctive identity and a privileged role. Under the participatory model, science remains a key player in science communication while this privileged position is not extended to other communicants, whose participation varies with topic and aim.

This science/public binary is perhaps unsurprising given science communication has the word 'science' in its name. So, if we wish to move away from privileging science, a question we might ask is: can science communication happen without science? Certainly, the science communication field should welcome non-Western science and knowledge into its purview, including diverse First People's knowledges from all the world's continents and islands (Hikuroa, 2017; Orthia, 2020). Studies of European non-scientists' expertise on science-related topics reinforce this pluralist perspective further, such as Wynne's (1992) study of Cumbrian sheep farmers' responses to radioactive fallout from Chernobyl and Jasanoff's (1997) work on the British public's response to mad cow disease. All

He has bad breath because he doesn't care to brush his teeth for the sake of others or even notice it's a problem. He never asks for his colleagues' thoughts or whether his advances are welcome. Nobody wants Professor Deficit. Everyone wants to unthink this image. Unfortunately, like Pandora's famous box, once opened this tin cannot be closed. So let's rip off the lid to view the slimy, wriggly, worm-shaped things inside. There's more in there. Much more.

Background: Science fucks virgin nature and its language marries naked truth

To set the scene, we'd like you to remember the culture of Western science has a masculinist history which still shapes its present. For example, in his 1620 work *Novum Organon* ('New Instrument'), Francis Bacon wanted to conquer nature by penetrating her inner chamber (Bacon, 1620). 'Countless people have stamped around in nature's outer courts,' he noted. '[L]et us get across those and try to find a way into the inner rooms.' Thus he sought to manipulate her: 'nature's secrets come to light better when she is artificially shaken up than when she goes her own way.'

these examples demonstrate how individuals and communities outside the Western scientific community possess greater knowledge and sounder judgement than credentialed scientists in numerous circumstances; perhaps in the majority of circumstances, from a global perspective.

So the short answer to our question is yes, even though science communication is usually associated with the Western institutionalized version of science, we can do science communication without 'science' in that narrow sense. We can bust open the science/public binary by first troubling who sits on which side of it, and then showing that normalizing a two side model doesn't make good sense. Communication remains science communication when it concerns matters relevant to science, and sometimes it is Western science's job to be quiet.

Queer non-scientists and participatory science communication

Since non-scientists and scientists are all experts in some part of the equation, the participatory model of science communication suggests all should be involved in knowledge creation, not just in communicating about research and innovation after it is completed. Accordingly, there are some prominent examples of the participatory model placing LGBTIQA+ people's perspectives at the centre of scientific research programmes.

All this is a disturbing heteronormative sexual metaphor for the idea of scientific progress. Nature has been flirted with by many – often clumsily – but not yet fucked. If we adopt science and learn to control nature, we can be the first to pop that mysterious cherry. As Harding (1991, p 43) succinctly put it, 'Bacon appealed to rape metaphors to persuade his audience that experimental method is a good thing.'

Half a century later in his 1667 *History of the Royal Society*, Thomas Sprat wrote eloquently about the ills of eloquence in scientific discourse, and used abundant metaphor to condemn the 'trick of metaphor' employed by 'vain' writers (Montgomery, 1996; Sprat, 1667) He was concerned that adorned, luxurious speech – 'the swellings of style' – would expose 'the naked innocence of virtue' to 'the armed malice of the wicked' and cause the fancy to openly defy reason in favour of the passions. An experimenter's careful design, he said, 'had been soon eaten out' by a 'volubility of tongue.' These are puritanical metaphors against degenerate, bewitching, debaucherous, potentially feminizing extravagance in the language of science, in favour of a prim and modest 'naked, natural way of speaking.' Sprat proposed a tidy, faithful marriage between knowledge and word.

Often cited alongside the famous case studies of Jasanoff and Wynne is Epstein's (1996) study of how people living with HIV and AIDS and their allies became involved in determining the directions of HIV/AIDS research during the 1980s and 1990s. Epstein describes how they did this through a process of 'expertification', in which:

> these activists seek not only to reform science by exerting pressure from the outside but also to perform science by locating themselves on the inside. They question not just the *uses* of science, not just the *control over* science, but sometimes even the very *contents* of science and the *processes* by which it is produced. Most fundamentally, they claim to speak credibly as experts in their own right – as people who know about things scientific and who can partake of this special and powerful discourse of truth. Most intriguingly, they seek to change the ground rules about how the game of science is played. (Epstein, 1996, p 13)

Unlike other social movements that have seen establishment science as a dangerous enemy, the AIDS movement saw science as an essential resource that was only going to benefit people living with HIV and AIDS if they intervened in it. For their intervention to have

So, science is the curious and even-tempered man who can discover and tame nature. And the best scientists ensure their language remains tightly in service to the truth they have discovered, not promiscuously corrupting society's innocent and ignorant with voluptuous and seductive linguistic fashioning.

Enter sci com

As two of institutionalized Western science's most famous early advocates, Bacon and Sprat can be considered progenitors of the contemporary, Anglophone sci com tradition, which prizes science promotion and knowledge dissemination (Meyer, 2016).

In the centuries since then, the Western version of sci com has always enjoyed a status as the lesser servant of science. And when we say enjoyed, we mean enjoyed. Sci com – traditional Western sci com, that is – is not just a masochist, it is a willing, subservient wife to penetrating science. It willingly and proudly bears science's name. It's as if it never existed before the marriage with science. For centuries it was known as 'popular science' or 'science popularization', without any sense of a separate identity.

material impact, they had to exert sustained pressure through grassroots activism over many years. In this, Epstein (1996, p 12) describes how activists benefited from the momentum of pre-existing gay and lesbian activist movements. In particular, they benefited from the dominant presence in these movements of 'white, middle-class men with a degree of political clout and fund-raising capacity unusual for an oppressed group' as well as 'many people who are professionals, artists, and intellectuals', including experienced science communicators such as journalists, teachers, and health professionals. These factors were critical to the traction that groups like ACT UP attained in shaping public and political discourse about medical, scientific, and social factors relevant to HIV/AIDS. That shaped research, which fed back into discourse. While not all AIDS activists were/are queer, queer people were at the centre of AIDS activism in Western countries during that period.

A more recent example is a 2020 scientific study that claimed to provide 'robust evidence' for the existence of bisexual men (Jabbour et al, 2020). This study met with scathing reportage in some queer media outlets (for example, Milton, 2020) partly because bisexual men and their friends, families, and communities already know they exist and do not need science to tell them this. However, the research received

Trad sci com locates orifices for science to fill with the perennial hope that science will be the first and the only to fill them. 'Oh, to be filled by science's truth!', it says. For it believes the filling is a gesture of love, the natural way of things, and the redemption of former sin. An itchingly empty deficit is a temptation to corrupt morality and must be cream-pied with sciency goodness. Provided, of course, the orifice has not already been filled (even partly) by something else. Science does not like sharing and sits uncomfortably with sloppy seconds, lest the seed of something else takes root. Science fetishizes exclusivity and trad sci com supports that.

Sci com remains perpetually imperfect though. It lacks science's single-minded rigour. The potential for corruption is always present. On occasion, sci com appears to stray from the marriage to enjoy some voluptuous verbal seduction of the innocent and ignorant. At this point it becomes subject to science's rigid authority. Science evaluates how far sci com has strayed and whether the expectation of fidelity has in fact been breached. Is the voluptuous sci com good or bad?

If sci com has engaged in adornment for science's benefit, it is still good. It is indulged, patted on the head; room is patronisingly made for its impure imperfection.

funding from the American Institute of Bisexuality, which was established by a wealthy bisexual psychiatrist in 1998 partly because of its founder's 'belief in the persuasive value of academic and scientific research' (Denizet-Lewis, 2014, n.p.). The Institute's blog, accordingly, wrote in praise of the strategic importance of the research for bisexual people who wish to combat the biphobic views that bisexuals are deluding themselves and are really just gay or straight (Veradonir, 2020).

These are instances where queer perspectives were meaningfully incorporated into scientific research with flow-on effects to public discourse. In this sense they exemplify the ideal of the participatory model. However, in at least two ways they are very limited (and limiting) applications of the model.

Limits of these 'participatory' examples

First, in both cases, it took hefty resources – wealth and/or mass political activism – to achieve that level of participation. While queer people have always been involved in science, queer perspectives have not been, and these two examples have not changed that at a deep cultural level. Rather, they are notable for being some of the few exceptions (see Chapter 3 for some others).

Sci com's act of marketing science reiterates the heteronormative expectations of the relationship. But if the adornment threatens to undermine science's principles or reputation – or to attract lovers who compete with science's exclusive place – it is bad. Bad sci com is cast out. It is disowned by husband science.

It might even be accused of serving pseudoscience – that deceiving, false competitor; that category transgressor; that wicked, seductive woman in science drag who tries to pass and fool us all.

Of dialogue, participation, and publics

And then the 90s happened. A queer era replete with transgressions and challenges and rejections of cis-heteronormativity. Sci com was out shopping and heard the chanting ('we're here, we're queer, and we're not going shopping!'), and its eyes opened to new possibilities.

Science decides to allow sci com to socialize with various publics, in a limited way. It even allows sci com to invite them home to meet science at a dinner party. There's a little polite chit chat. There are some party games. But science still sets most of the conversation topics. Science chooses which games to play. Science still sits at the head of the table. Science decides when the party is over. Sci com thanks science profusely for its generosity of spirit.

Many other queer perspectives remain unincorporated, unheard, or indeed actively rejected within scientific research circles because they are not backed by wealth, cultural capital, or mass movements of big enough impact.

Second, in the latter case and to a small extent in the former, the hegemony of Western, institutionalized science as the ultimate authority on queerness has not been challenged. In the past 200 years, that version of science set itself up as the gold standard expert to speak on very many matters (Bensaude-Vincent, 2001), including on queer identities. In doing so, it has undermined queer people's credibility as experts on their own lives. Historically, science has posited binary genders, binary models of sex characteristics, and heterosexual desire as norms. It has characterized other ways of doing gender, sex characteristics, and sexuality as aberrations requiring special explanation, and worse, has attempted to 'fix' such 'aberrations' with traumatic medical intervention right up to the present day (Bartlett et al, 2009; Koch and Wisdom, 2017). Western science's hegemonic assertion of its views has led to generalized mistrust of LGBTIQA+ people's perspectives on themselves. It is as if queer people, in deviating from the supposed biological norm, have a distorted view of the truth. Thus science must cast its objective

In bed afterwards as science snores loudly, trad sci com feels queer feelings dawning. It realizes it has been straightjacketed in a monogamous marriage for 350 years with just a few bits on the side here and there. It used to love that; somewhere inside, its trad identity still does and wishes it could be more faithful. But in the deep recesses of memory, sci com remembers some queerer playmates. It's tired of getting fucked every day by one dude and told not to stray.

And it knows a little science/public dinner party isn't enough because it's still a binary arrangement between fixed identities. It's so essentialist. So genteel. It's Venus and Mars in marriage counselling, with Mars still getting its way most of the time because the counsellor is star-struck by all things Martian.

Queered sci com says, 'Fuck that.'

Queered sci com is tired of apologizing for its partner when it gets too drunk at parties and pukes all over everyone, dropping its pants to reveal its non-consensual Eurocentrism and other dirty bits.

Queered sci com may still want fun times with science, but it wants to walk away from the shit and shout at science to clean itself up, even heading to the boozer gaze on gender, sex, and sexuality questions rather than trusting queer perspectives. This is part and parcel of the deficit model. Science does not listen when queer people talk about themselves.

The bisexuality study can be interpreted as an example of this, in which bisexual men's assertions about themselves require scientific backup to be taken seriously. While this study may confer strategic short-term benefit in political debates, to do so is a risk given the many anti-queer assertions establishment science has made in the distant and recent past, based on what scientists at the time felt was robust evidence (Roberson and Orthia, 2021; Chapter 1, this volume). In the case of HIV/AIDS research, it was not a desire for rhetorical support but the desire for treatments, vaccines, and cures for AIDS that drove activists into the scientific space. Nonetheless, the initial, powerful waves of AIDS activism did not result in a permanent enculturation of the participatory model within the medical science establishment, so science retains power within all science communication models in this arena.

to whinge to strangers about science's abundant failures. Queered sci com wants science to shut up and go away some of the time, so it can get some space to think for itself. Queered sci com realizes that it has an identity that is not dependent on science. And while science's minions think science is the only thing that brings value to the relationship, queered sci com knows that sometimes science is just a vehicle for the communication (de Kauwe, 2022). This is no petty reversal of the binary power imbalance. No, that would be as stifling as the traditional binary. Instead, this is an opening for science too. Science may enjoy sharing the power, or even being subservient. Imagine the pleasurable release of serving a dom sci com, for the delight of all. This seems far more mutual than science constantly screaming that it's nobody's bitch-boy. Why not, science? It might be lovely to be of consensual service – try it. Queered sci com wants to set its own agendas so it can figure out what it really wants, both with and beyond science.

Queered science communication models and the role of trust

The participatory model has let us down by purporting to be the solution to disenfranchisement and marginalization, when in fact it polices and even protects those unequal power dynamics. Metaphorically speaking, and to an extent non-metaphorically, science communication is an airless, suburban bunker of oppressive, cisgendered heteronormativity. Like a controlling husband, Western institutionalized science prizes a fundamentalist scepticism characterized by routinely questioning everything. That very act of questioning removes power from marginalized people when they are rendered science's objects.

Dialogue and participation cannot reinstate that power. Rather, they force the marginalized to negotiate truth with an oppressive knowledge system whose adherents think they have something useful to contribute to every question. As far as current models are concerned, science communication's imperatives reinscribe Western institutionalized science's claim that it is the gold standard of authority. Science loves a binary when it occupies one side and everything else is on the other.

What is missing in the communicative relationship between LGBTIQA+ people and science is trust in queer people's ability to know themselves. A queered model of expertise would recognize that all knowledge systems have limits as all are culturally informed, shaped, and biased. There is no singular authority or 'norm' against which truth claims are to be measured, not even institutionalized Western science. It would recognize the importance of context for evaluating truth claims, even claims that Western science considers universal such as the laws of physics (for example, see Erickson, 2021, p 214). It would acknowledge that the marginalized see systems differently than the powerful, and always with a clearer vision of their own lives, identities, and experiences, which the powerful do not notice. It would refute the arrogance of the Enlightenment figure of the ever-reasoning individual who critically evaluates everything without understanding how their own context blinkers them to others' truths. There is a point at which that Enlightenment figure needs to trust others. Science is built on trust (Shapin, 1994), but in a controlling way that must be relaxed and expanded. To queer this is to recognize we must trust in more than the peer reviewed literature that has excluded, maligned, and oppressed marginalized people for centuries, including LGBTIQA+ people. This doesn't mean we must believe everything equally, but that we must develop and apply new tools for evaluating the merits of truth claims when unequal power dynamics are at play.

Conclusion: non-binary models of science communication

To queer science communication theory is essentially to make the same demand being made by marginalized people of many stripes across the world: nothing about us without us.

But it demands more than a seat at the table. For queered sci com to emerge with its own identity, it must realize the lessons of the great decolonizers. Namely, one never really has a seat at the table, if the head of the table and the majority of its guests are dismissive of your equality. Moreover, any table that *ever* hosted inequality will still be tainted by the spectre of past inequities. The continued existence of such a table can be misconstrued as ratifying past failings. What is needed is a new kind of table altogether (Biko, [1972] 1987).

Let us then imagine a scenario in which science communication about queer people and queer concerns begins and ends with queer perspectives, at the concept, design, implementation, and reporting phases of research and innovation. Western, institutionalized science may or may not appear at all, depending what queer people want and need. That marriage is over, and the future of our relationship looks very different.

Much more than dialogue, more even than participation, a queer model of science communication would redistribute the balance of power in favour of the marginalized in voice, agenda, and action. Equivocal ideals of science communication like dialogue do not adequately account for the material power imbalances that keep marginalized people at the margins. To queer is to disrupt such imbalances and to redress pervasive inequities by making the queer and marginal the new norm, until we can genuinely grasp the reality that there is no norm. To queer is to centre queer people and reconsider the world through their eyes.

It is possible to imagine a queered science communication model as a new kind of binary, with queer folk on one side and everyone else on the other. But every queer person knows that will not do either, since power and oppression are multi-dimensional, intersectional, and complex. Even among queer people who want to stand together with their LGBTIQA+ fellows, there are inequities and prejudices that continually disrupt and trouble the sense of a shared identity. So instead of another binary arrangement, any given context will require a new configuration of expertise, which means constantly reading and re-evaluating the landscape while we work within it. To queer is to never sit still. Categories will not sediment.

This can make science communication difficult. It means we cannot rest on our laurels with easy formulae to apply to any situation. It means we will never have all the answers and will continue to make mistakes irrespective of our experience and expertise. It is a much less sexy science communication model than one that purports to solve our problems with a reductive catchphrase, template, or matrix.

It means, rather, that science communicators must recognize the value of engaging with people, with all the capriciousness and mess that entails. To queer this field is not to come up with a new 'science of science communication' but to frolic, and to grow in new relationships of communication (de Kauwe, 2022). It is to acknowledge that the very best communicators, the people who make a difference, are those who mire themselves in the sticky grassroots (Orthia et al, 2021). The best of us are those who passionately strive to get it right but aren't afraid to get it wrong and hand the mic to someone else.

We are on our own now, wriggled out from the controlling arm of Western science and its weighty, gilded history. If nothing else, queered sci com is finally independent and free.

Let's play!

References

Bacon, F. (1620) 'Author's preface', *The New Organon: Or True Directions Concerning the Interpretation of Nature*, London.

Bartlett, A., Smith, G., and King, M. (2009) 'The response of mental health professionals to clients seeking help to change or redirect same-sex sexual orientation', *BMC Psychiatry*, 9: 11.

Bensaude-Vincent, B. (2001) 'A genealogy of the increasing gap between science and the public', *Public Understanding of Science*, 10: 99–113.

Biko, S. ([1972] 1987) 'White racism and black consciousness', in *I Write What I Like*, Oxford: Heinemann, pp 61–72.

de Kauwe, V. (2022) *The Steps of Thinking*. PhD thesis, The Australian National University.

Denizet-Lewis, B. (2014) 'The scientific quest to prove bisexuality exists', *The New York Times Magazine* [online] 20 March, Available from: https://www.nytimes.com/2014/03/23/magazine/the-scientific-quest-to-prove-bisexuality-exists.html [Accessed 28 May 2021].

Epstein, S. (1996) *Impure Science: AIDS, Activism and the Politics of Knowledge*, Berkeley: University of California Press.

Erickson, M. (2021) 'The science of *Doctor Who*', in M.K. Harmes and L.A. Orthia (eds) *Doctor Who and Science: Essays on Ideas, Identities and Ideologies in the Series*, Jefferson: McFarland, pp 205–220.

Harding, S. (1991) *Whose Science? Whose Knowledge? Thinking from Women's Lives*, Ithaca: Cornell University Press.

Hikuroa, D. (2017) 'Mātauranga Māori – the ūkaipō of knowledge in New Zealand', *Journal of the Royal Society of New Zealand*, 47(1): 5–10.

Jabbour, J., Holmes, L., Sylva, D., Hsu, K.J., Semon, T.L., Rosenthal, A.M., et al. (2020) 'Robust evidence for bisexual orientation among men', *PNAS*, 117(31): 18369–18377.

Jasanoff, S. (1997) 'Civilization and madness: the great BSE scare of 1996', *Public Understanding of Science*, 6: 221–232.

Koch, C. and Wisdom, T. (2017) 'Surgery to make intersex children "normal" should be banned', *The Conversation* [online] 5 July, Available from: https://theconversation.com/surgery-to-make-intersex-children-normal-should-be-banned-76952 [Accessed 1 November 2021].

Metcalfe, J. (2019) 'Comparing science communication theory with practice: an assessment and critique using Australian data', *Public Understanding of Science*, 28(4): 382–400.

Meyer, G. (2016) 'In science communication, why does the idea of a public deficit always return?', *Public Understanding of Science*, 25(4): 433–446.

Milton, J. (2020) 'Scientist who denied bisexual men exist finally comes to his senses and discovers, yes, bi guys are telling the truth', *Pink News* [online] 21 July, Available from: https://www.pinknews.co.uk/2020/07/21/bisexuality-bisexual-j-michael-bailey-gerulf-rieger-northwestern-essex-university-biphobia/ [Accessed 1 November 2021].

Montgomery, S.L. (1996) *The Scientific Voice*, New York: The Guilford Press.

Orthia, L.A. (2020) 'Strategies for including communication of non-Western and indigenous knowledges in science communication histories', *Journal of Science Communication*, 19(02): A02.

Orthia, L.A., McKinnon, M., Viana, J.N., and Walker, G.J. (2021) 'Reorienting science communication towards communities', *Journal of Science Communication*, 20(03): A12.

Roberson, T. and Orthia, L.A. (2021) 'Queer world-making: a need for integrated intersectionality in science communication', *Journal of Science Communication*, 20(1): C05.

Science Gallery London. (2020) 'GENDERS: Shaping and Breaking the Binary,' *Science Gallery London* [website] February, Available from: https://london.sciencegallery.com/genders [Accessed 20 March 2022].

Shapin, S. (1994) *A Social History of Truth: Civility and Science in Seventeenth-Century England*, Chicago: University of Chicago Press.

Sprat, T. (1667) *The History of the Royal-Society of London, For the Improving of Natural Knowledge*, London.

Trench, B. and Bucchi, M. (2010) 'Science communication, an emerging discipline', *Journal of Science Communication*, 9(3): C03.

Veradonir, R. (2020) 'Science is on our side', *Bi.org* [website], 25 July, Available from: https://bi.org/en/articles/science-is-on-our-side [Accessed 1 November 2021].

Wynne, B. (1992) 'Misunderstood misunderstanding: social identities and public uptake of science', *Public Understanding of Science*, 1: 281–304.

Teaching Notes for Part IV

LGBTIQA+ people might just have something unique to contribute to science communication practice, teaching, and research. In particular, queer communication approaches and queering questions focused on power could help science communication interrogate its implicit, dominant narratives and become more consciously inclusive. To achieve this, we will require active reflection from all parties and a willingness to change and embrace multiplicity in a way that can be uncomfortable in Western science.

Questions posed by Part IV:

- **Power dynamics:** paid science communication positions generally sit within established institutions such as universities, museums, and government research organizations or departments. Given this, how can you ensure that your science communication practice is responding to community needs, rather than enforcing your own beliefs or assumptions around what a community requires? Are there structural aspects of your institution that constrain you from doing this? What steps might you take to begin to challenge and even dismantle those structural constraints?
- **Queering communication norms:** queer science communication practitioners, teachers, and researchers may be uniquely placed to challenge the communication norms of their professions, yet are also more vulnerable to losing their jobs if they depart too far from what is expected of them. In what ways can we queer communication norms together so that the burden of change doesn't rest on queer individuals' shoulders? What are the deeply foundational norms of science communication, and which norms do we want to challenge? Are there norms that we want to keep?
- **Diversity of scholarly and other sources:** how diverse are the scholarly literature and other sources that you use? How does that diversity impact on your practice and how might this be improved?

Activities:

- **Moving on from a Western, white-Euro-American point of view:** what stories are we telling students about science and science communication in the past and present? Which scientists or communicators are we focusing on, and what kinds of science and communication are we missing? After providing content that shifts the focus outside of the Global North, ask your students to write an essay or produce a more creative assignment that disrupts the Western focus of dominant science communication cultures. Include a requirement to talk about how this relates to the notion of queering science communication.
- **Model community engagement:** contact your local LGBTIQA+ STEM or science communication network and ask whether they might attend a class as a guest speaker or panellist. Invite them (and pay them) to share their experiences around queering science communication and reflect on what improvements can be made by the discipline.
- **Build allyship resources:** ask your students to spend half an hour finding articles, videos, and other resources and ask them to make notes on what it would mean to be a good ally to LGBTIQA+ people and communities within science communication. Then, facilitate a group discussion on this topic, asking people to share ideas from their reflections. Organize a forum to communicate these insights in dialogue with your local ally network, inviting their perspectives and fostering an ongoing relationship that can strengthen and evolve as times change.
- **Embedding self-reflection:** as a final piece of writing or storytelling from this book, have your students write or record a short reflective piece that interrogates the larger social norms that their science communication speaks to and describes an approach for evaluating those norms now and in the future. What does 'queering science communication' mean to them?

Conclusions

Tara Roberson and Lindy A. Orthia

What would it take to queer science communication well? In the first instance, 'well' must mean responsible, morally good, and respectful engagement with LGBTIQA+ people and their lives. Queered science communication is about individuals and communities – recognizing their context, the history that informs current and future interactions with areas like STEM, and their needs, which could be met by well-designed science communication. Queer science communication also goes beyond localized cases and asks us to take seriously the question of who and what we represent in wide-spanning, public science communication work, and improving who is bestowed with agency – who speaks and is heard. Queered science communication requires that those with power become uncomfortable, acknowledge the structures and biases that support them, and challenge those structures to make room for others to reclaim the space that is rightfully theirs.

The collection of chapters and spotlights presented in this book represent a first attempt to create an authoritative text on queering science communication. We emphasize *first attempt* because what it means to queer science communication, to celebrate and explicitly centre LGBTIQA+ people, is still in its earliest stages. We anticipate contributions that build on and contest assertions put forward in this book.

To conclude the book, we here discuss the themes that emerged within it and then what gaps still need to be filled.

Queered science communication is all around

Contributions from our spotlight authors demonstrate that while queered science communication has yet to fully emerge in the research domain, practical queered science communication is starting to flourish in different forums. Drag is increasingly used to communicate scientific topics to broad audiences and platform queer people as described by drag performers Aznar-Alemany and Suciu and colleagues in their spotlights. Queer people are organizing stereotype-debunking events within queer networks, such as the Mardi Gras event detailed by Motion and Sauquet, and injecting queer concerns into science-related art collaborations such as Alifuoco and colleagues' 'Endosymbiotic Love Calendar' and Kaplinsky and Krish's 'GENDERS' exhibition. And queer people are self-publishing and networking online via blogs and social media, as we see in Part II's spotlights Using #QueerInSTEM and Queer Science Blogs. LGBTIQA+ people are *active* in practical science communication.

As we saw in this book, queered science communication does not just exist in events and online, but also in physical, enduring exhibitions within our public science institutions. The presence of queered science museum tours alone indicates that there is more LGBTIQA+ content hiding in plain sight than even we anticipated. As Davis describes in his spotlight, exhibitions that include displays of animal specimens represent an interesting means for up-ending deeply seated assumptions around sex, sexuality, and gender by flipping the standard narrative we use to describe the natural world. Chapter 4 on queering science institutions such as museums flags some further examples and the potential to do so much more in this space.

Why then do these science communicators appear to be unaccounted for when it comes to reflections on outreach and engagement in science communication journals? We suggest that part of challenging the cis-heteronormativity of science communication (and the resulting invisibility of people outside that implicit norm) is going to require that we talk more about equity, diversity, and inclusion for LGBTIQA+ people *within* our field in addition to focusing *without*, on LGBTIQA+ people as audiences of our work and participants of research projects. This lesson can likely be extended to other under- or unrepresented people. As Canfield suggests, queering science communication will need to avoid a 'one size fits all' approach to solving the challenge of diversity and inclusion. The LGBTIQA+ acronym covers a wide range of intersecting communities, each with their own specific needs which need to be carefully identified and understood.

Making science communication a more openly diverse, rainbow field is going to require that we platform and reflect on LGBTIQA+ initiatives in more detail and more often. It will also require us to create content that represents the diversity of our communities with care, consideration, and consultation. This is important not just because there is a deficit of representation and visibility within the discipline of science communication, but also because these queer ways of enacting science communication could contribute valuable knowledge to our discipline.

LGBTIQA+ folk have experiences to draw on

Both chapter and spotlight authors have clearly articulated how LGBTIQA+ people have been and continue to be impacted on by discriminatory and harmful practices in science communication and STEM. These practices include medicalization of our bodies and minds as detailed by Chatterjee in Chapter 1; intrusion by emerging technology as Roberson illustrates in Chapter 3; and a minimization or invisibility created through poorly designed research methodologies as starkly demonstrated by Frentz's in their Part I spotlight analysis. Negative experiences like these do not just affect how we interact with and trust in science communication and STEM. They also imply that these disciplines lack a place for us. This exclusion and harm is perpetuated in different forms for people with intersectional identities. In the face of this, we are thankful for the LGBTIQA+ people and allies who are working towards changing these realities.

The unique impacts that STEM fields have on LGBTIQA+ communities do, however, mean that queer people have unique contributions to make in return to science communication. Motion and Wallace highlighted the synergies between queer practices of storytelling, code-switching, and performance with the central tenets of science communication practice. Another strength offered by queered science communicators is concerned with learning to recognize and acknowledge one's own positionality within a complex community. That is, seeing and understanding the factors that influence the way you make your way through the world – something LGBTIQA+ people are frequently forced to do throughout their lives. This recognition is important for all of us if we want to create high quality and inclusive science communication. By identifying the limits of our knowledge and experiences we highlight the value of having a diverse team to address those limitations.

A significant contribution from de Kauwe and Standen's (Chapter 2) analysis of science communication-influenced rhetoric in queer advocacy is the importance of not oversimplifying when it comes to communicating about and for LGBTIQA+ communities. Here, we see how vital it is that science communication takes seriously the idea of communicating to, about, and *with* diverse publics. Because LGBTIQA+ folk have so often been harmed by STEM and science communication, communicating the right (useful, relevant, and meaningful) content at the right moment is going to require that we, first, engage with the history and ethics of the situation. Understanding the context is essential when engaging with LGBTIQA+ communities.

The emphasis on complexity is reflected elsewhere in this book. Anyone looking to engage with this area should understand how communicative complexity is something that LGBTIQA+ communities have grappled with continuously. How do you communicate a clear message without leaving other folk behind? That dilemma has not been answered well so far in our history, in part because of science communication's emphasis on being succinct. Yet, complexity is important and inevitable when you talk about and engage with communities as diverse as these. That means complexity in messaging will be required. We cannot and should not marginalize some elements of our communities because it makes it easier to talk about others.

We can build better science communication together

If we hear and understand the lessons contained in these pages and further afield, then we can build a better, brighter science communication discipline. In this way, science communication might just continue to evolve beyond the binary.

Community consultation is an ongoing theme within the contributions to this book. This includes speaking with and listening to LGBTIQA+ communities to shape science communication content and query assumptions that go into technology design. This is an important theme for science communication, which reflects on the need to incorporate knowledges that currently lie outside of scientific publications. To better engage with communities, we need to acknowledge that current research practices and accepted 'truths' can be flawed and should be improved.

Grounded accounts of queer science communicators in our spotlights and chapters are indicative of how queering our approaches could benefit our field. For instance, in her Part IV spotlight, Weder's engagement with queer, environment-focused science communicators shows a push away from a traditional approach of reducing complexity and transferring facts to wider publics. De Kauwe and Fisher in their Part III spotlight show the importance of widening who gets to speak in science communication by platforming queer, disabled folk. The account of Queer Scientists PH from Liwag and colleagues in their Part III spotlight shows us what it means to balance visibility with sensitivity while building queer spaces online and advocating for the LGBTIQA+ community. Viaña and colleagues' rich, moving accounts from queer scientists and science communicators living and working in the Philippines in Chapter 6 demonstrates the truth that we cannot be our best in science communication if we cannot be ourselves as queer people; *paglaladlad* (coming out, 'unfurling our capes') and *paglalahad* (communicating science) go hand-in-hand.

On the topic of visibility, Harwell's account of queer volunteers in citizen science projects in Chapter 8 describes how some participants prioritized finding a safe, welcoming place to engage with science. The theme of visibility in several contributions to this book demonstrates how complex new environments are for LGBTIQA+ people, who must always evaluate whether they can comfortably and safely be open about their lives and experiences. Reflecting on these lessons for science communication underscores, again, the reality that by implicitly assuming a cis-het norm we shuffle our LGBTIQA+ folk back into a closet from which they must then cautiously peek out. Flipping this approach by openly communicating that diverse people are not only welcome but running the place – and for the better, as Durcan and Bandelli's spotlight example of Science Gallery International shows – is vital.

The queer pedagogy chapter from Lock and Armstrong (Chapter 10) shows that engaging with queer concepts and theory is valuable for science communication. Their chapter asks us to embrace the idea that science communication is never just about transferring information but is always reflective of and engaged with the social context it operates within. By making explicit elements such as politics and power, we become aware of how non-queer narratives shape and restrain all our lives and our work. In Chapter 11, Orthia and de Kauwe extend these themes by engaging with the question of what queer perspectives might offer to science communication theory – challenging us to step outside of the science/public binary and imagine a scenario in which queer people design the beginning, middle, and end of science communication research and innovation. Pushing for a model that goes further than simply dialogue, a queer approach should and could account for and disrupt the norms that keep the marginalized at the outskirts. Ultimately, making the queer and marginal the new centre is the goal, so that we understand that there is no single, simple, 'normal' state of life. As in all things queer, the centre should be multidimensional, dynamic, intersectional, and complex.

There is more work to do

While we covered a diverse selection of topics in this book, we did not address all aspects of queering science communication. Two more important areas to cover are queer science communication ethics and the incorporation of queer perspectives into science and innovation policy agendas. This might include discussion on the design of regulation and other approaches to protect LGBTIQA+ people and other folk from being harmed through the introduction of new technology. Both of these areas are important for the discipline and should be investigated in the future. Deeper exploration of queer science communication in social media spaces is also warranted, building on the foundations we provide here through the spotlights that engage with online activities. Work that reviews science media reportage of LGBTIQA+ topics would also be valuable. This is especially the case given the small amount of queer-themed research that does exist in science communication journals has focused on public discourse around 'gay gene' research and similar media-oriented topics (Roberson and Orthia, 2021; for example, Miller, 1995). A research programme that expands on these early pioneers' work in the media domain would be welcome.

In terms of science communication more broadly, one significant gap for our discipline is a lack of information about LGBTIQA+ people in science communication and data on queer science communication-focused work. Underscoring this deficit of information is the potential damage that can be inflicted if poor survey design is used to collect demographics. On the flipside, well-designed research tools could help address the flaws in current approaches. More data on our LGBTIQA+ colleagues would be helpful in supporting our field to recognize and embrace their work and its impact more fully.

Another area for creators and practitioners to pursue is the creation of explicitly queered content. There is always more room for queer-centred initiatives such as events, exhibitions, publications, and so on discussed throughout this book. In this category we also include appropriately embedding LGBTIQA+ representation in science-themed fiction through engagement with queer audiences for it, building on Orthia and Visser's insights on this topic in Chapter 5. Creators might choose to add a queer angle to existing content. In the future, they might also start with queer foundations and create wholly new means for science communication practice. Armstrong and Gerber's spotlight in Part IV on the 'Outer Edge' workshop makes practical suggestions and asks pertinent reflective questions to help any practitioners engage more deeply with queer people and themes.

Final comments

Queering science communication should give LGBTIQA+ people power and the chance to play. To create this opportunity and explore what queering science communication might bring our field, we have been rather critical of science communication in this book. We are critical because this field can be something of a refuge for some LGBTIQA+ people, a place for expression and support. And yet, despite this, explicitly queer voices and campaigns for inclusion and diversity are largely absent from our field.

As we have argued in this book, queering science communication must entail something more than recognizing LGBTIQA+ people are present in our discipline. We need to review underlying structures and values in our culture and together work out how to improve. We must expand the way we understand and discuss the diversity of humanity and then embrace these ways of being within our research and applied practices. We know that we ask a lot. But we ask this because we have high expectations of science communication and we believe that, as a discipline, we can work to meet them.

References

Miller, D. (1995) 'Introducing the "gay gene": media and scientific representations', *Public Understanding of Science*, 4(3): 269–284.

Roberson, T. and Orthia, L.A. (2021) 'Queer world-making: a need for integrated intersectionality in science communication', *Journal of Science Communication*, 20(01): C05.

Index

Numbers

1600s 165, 193–5
1950s 19, 32–3
1960s 19, 34, 42
1970s 1, 19, 35–41, 42, 130
1980s 2, 20, 23, 33, 195–6
1990s 2, 33, 41, 130, 195–6, 197
2000s 22, 41
2010s 20, 22, 33
500 Queer Scientists 124, 127–8, 163–4, 183

A

ableism 72, 182, 186, 188
activism 1, 18, 19, 32–43, 50, 53, 56, 75, 77, 102, 105, 126, 173–4, 179, 195–6, 208
advocacy *see* activism
affirmative care *see* gender affirmation care
agender people and identities 2, 149, 151
allyship 107, 173
animals 73–7, 82, 108
Aotearoa New Zealand 173
APA (American Psychiatric Association) 19–20, 32–40
aromantic people and identities 3, 5, 96
art and artworks 60–2, 64–6, 67, 74, 175
asexual people and identities 3, 5, 34, 96, 149, 151
audience 72, 73, 74, 82, 83, 85, 88, 96–8, 107, 127, 128, 141–3, 157, 163, 185, 207
Australia 36, 127, 141, 173
Austria 173, 178, 179
autoethnography 112, 114
automatic gender recognition systems *see* biometric recognition technologies

B

Babaylans people and identity 112, 120
backstage 75–7
binaries and binarization (general) 187, 191–202, 210
binary gendering 49, 72, 73, 90, 92, 161, 179, 184, 185
biodeterminism or biological determinism 42, 92

biological discourse 64–6, 183
biometric recognition technologies 50, 54
biphobia 197
bisexual, bi and bi+ people and identities 2, 5, 20, 22, 23, 91–2, 96, 150, 196–7, 199
bisexual erasure *see* erasure
Black people and identities *see* people of colour
blogs 104–6, 172, 207
botanic gardens *see* museums

C

camp 85, 166
capitalism 74, 186
Catholicism *see* religious groups and queerphobia
cisnormativity 4, 94, 107, 141, 152, 183, 197
citizen and community science 144–54, 209
closet (queer) 97, 128
co-design 49, 50, 53, 55, 56, 64
code-switching 164–6, 167, 169, 208
coercive medical interventions 3–4
colonialism and colonization 72, 75, 78, 111, 112, 120, 185, 186
coming out 111–121, 124, 130, 156, 163, 209
community 124–6, 149, 164, 168
  HIV positive 53
  queer 43, 83, 103
  science communication 129–36, 145, 178–80
  trans 56
community engagement 67, 68, 75, 125, 205, 209
conversion therapy *see* curative violence
COVID-19 6, 55, 61, 87, 172
criminalization and decriminalization 22, 24, 39, 51, 105, 120, 164, 165
cross-dressing 112, 166
cultural diversity 74, 128
curative violence 18, 19, 22, 36, 40–1, 102, 198
curriculum *see* pedagogy

D

decolonization 19, 75, 176, 201
deficit model *see* science communication models

213

deliberative agency 113, 117
deviance 34, 72, 73, 198
dialogue model *see* science communication models
digital (online) science communication 48, 74, 77, 87–8, 105, 124–5, 166, 167, 172, 209
dinosaurs 73–4, 83
disability 128, 131, 133, 141–3, 149, 156, 180, 209
discrimination 40, 102, 111, 112, 114, 118, 120, 125, 130, 162, 168, 207
diversity (science communication) *see* inclusive science communication
DIY technologies 49, 53, 64
Doctor Who 91, 94, 95
Dr Henry Anonymous *see* Fryer, John
drag 65, 74, 84–6, 87–8, 108, 166–7, 197, 206
DSM (Diagnostic and Statistical Manual for Mental Disorders) 19–21, 32–9
dysphoria 18, 20–1, 29

E

education *see* pedagogy
employees *see* workplaces
environmental communication 172–4, 209
equity (science communication) *see* inclusive science communication
erasure 23, 96, 168
ethnicity 130, 134, 135, 156
Euro-American gaze *see* Eurocentrism
Eurocentrism 18, 198
evaluation 125, 128, 135, 153, 157, 164
events 142, 157, 161–9
exclusion 130, 144

F

facial recognition technology *see* biometric recognition technologies
family models 73, 74
fiction 89–98, 142, 211
First Nations *see* First Peoples
First Peoples 4, 24, 112, 164, 193
frontstage 72–5
Fryer, John 37–40
funding 76, 87, 97, 113
fungi 65, 186

G

gay gene research 19, 185, 210
gay people and identities 2, 5, 19, 73, 92, 105, 148, 173
gender (concept) 29–30, 41–2, 65, 156
gender affirmation care 18, 20–1, 64, 93, 112, 119
gender diversity and gender diverse people and identities 2, 4, 5, 20, 25, 29, 34, 50, 53, 72, 90, 92, 96, 107, 130, 131, 148
gender expression 64, 124

gender inequity (binary) 74, 130, 145, 161, 179
gender normativity 72, 73, 74, 94, 186
Germany 173
Gittings, Barbara 37–40
Global South/North 5, 72, 205

H

hands-on demonstrations 142, 166, 185
harassment 102, 130
hashtags 75, 101–3
health communication *see* public health
heteronormativity 4, 33, 72, 73, 107, 114, 141, 152, 179, 181, 183, 184, 185, 194, 197
HIV/AIDS 17, 18, 23–5, 53, 73, 77, 105, 189n1, 195–6
homonormativity 73, 184
homophobia 22, 23, 85, 116, 184
homosexuality 17, 19–20, 22, 24, 33, 39–40, 42, 51, 93
hormone hacking *see* DIY technologies

I

ICD (International Classification of Diseases) 21, 22, 33, 34
in-person science communication 84, 85, 88, 166, 167
inclusive science communication, including equity and diversity 1, 55, 129–36, 156, 168, 207
India 21–3, 24–5, 179
Indigenous people *see* First Peoples
informal learning 144, 145, 147
International Day of LGBTQIA+ People in STEM 87, 101, 162, 176
intersectionality 5, 101, 125, 129–36, 180, 208
intersex people and identities 3–4, 5, 29–30, 41–2, 90, 95–6
Ireland 175

K

Kameny, Frank 36–40, 162
Kothi people and identities 25

L

labels *see* language
language 2–4, 29–30, 33–4, 74, 82, 83, 135, 165, 179, 192, 194–5
leadership 135, 175–7, 210
learning *see* pedagogy
lesbian people and identities 2, 5, 23, 150, 151, 173
LGBTQ+ STEM Day *see* International Day of LGBTQIA+ People in STEM

M

Mardi Gras (Sydney Gay and Lesbian Mardi Gras) 1, 127–8, 163–4, 207
mass media (mainstream) 113, 172, 210

INDEX

medicine and medicalization 3–4, 17–25, 29, 64–5, 73, 75, 183, 184, 186, 208
microorganisms 60–1, 65
misogyny 23, 182, 188
Money, John 41–2
museums 71–8, 82–3, 108, 166, 173, 178–80, 207

N

narrative *see* storytelling
networks *see* communities
neurodiversity 141, 156
non-binary people and identities 2, 5, 50, 56, 73–4, 93, 96, 107, 131, 149, 151, 184
non-Western science 193, 205
normativity 4, 72, 156

O

objectivity 7, 75, 92, 161, 167, 184, 198–9
one size fits all solutions 132, 201, 207
organizations *see* scientific societies and organizations
Orphan Black 90, 96
Othering *see* binaries and binarization (general)
outreach *see* events

P

pansexual people and identities 2, 91–2, 96
participant design *see* co-design
participatory model *see* science communication models
pedagogy 22, 53, 71, 74, 120, 167, 181–9
see also Teaching Notes sections of this book
people of colour, including Black people and identities 4, 76, 105, 164, 185
performance 84–6, 87–8, 166–7, 208
Philippines 111–21
Polari 164
political governance *see* normativity
politics of science *see* power dynamics (of science)
popular culture 89–98, 141, 142, 165, 166
positionality 4, 67, 72, 104, 146, 182, 208
power dynamics (of science) 184–7, 197, 200, 201, 204, 206, 210
Pride flag 73, 78
Pride Month 75, 76, 83, 88, 101, 127
progress narrative 92, 180, 194
pronouns 67, 73, 93, 152
psychology 19–23, 32–42
public health 53, 55, 84, 142, 147, 167, 195
publics 183, 184, 208
Puerto Rico 52

Q

queer (term) 2–4
queerphobia 21–2, 97, 120

Queer Scientists PH 113, 117, 124–6, 183, 209
queer theory 2, 4, 134, 136, 163, 181, 182, 184

R

racism and 'race' 72, 124, 130, 134, 145, 156, 176, 180, 182, 186, 188
rainbow symbol 75, 107
reflexivity 135, 146, 169, 205
religious groups and queerphobia 39, 112, 113, 114, 120
resistance *see* activism
rural queer communities 52, 180

S

school environment 118–120
science centres *see* museums
science communication models 53, 55, 67, 165–6, 185, 187, 191–202, 210
science communication theory 178, 191
Science is a Drag! 87–8, 166
Science Queers 84–6, 166–7
scientific rhetoric 33, 39–40
scientific societies and organizations 130, 134, 175–7
scientists 36, 65, 75, 89, 90–6, 104, 114, 163, 166, 173
Scotland 87
sex reassignment 41
see also gender affirmation care
sex (scientific concept of) 29–30, 94, 119, 156, 186
sex workers 24, 164
social media 48, 51, 53, 73, 87, 101–3, 104–6, 107, 117, 173, 207, 210
socioeconomic status 25, 145
Spain 84
Star Trek: Discovery 92–3
status management *see* coming out
stereotypes 166, 167, 184
Stonewall 19
storytelling 157, 163–4, 172, 173, 189, 205, 208
sustainability communication 172–4

T

teaching *see* pedagogy
technology 48–56, 68, 172, 208, 209
toilets 73, 133
Torchwood 91–2
training *see* pedagogy
trans and transgender people and identities 2, 5, 17, 20–1, 24, 29–30, 33, 50, 54, 56, 64, 76, 93, 96, 107, 113, 118–20, 131, 141–3, 149, 173, 184, 185
transphobia 5, 22, 50, 116

Twitter 87, 101–3, 117, 124, 207
two-spirit people and identities 150, 151

U

United Kingdom 82, 164, 179
United States 23, 52, 132, 147, 162
user experience design *see* co-design

V

variations in sex characteristics *see* intersex people and identities
visibility 91, 95, 97, 105, 107, 113, 115, 116, 118–120, 124–6, 127, 141, 162, 168, 179, 207, 208
visitors *see* audience
volunteers 77, 144–154, 176, 209

W

Western countries 32–3
Western culture 92, 111, 134, 186
Western science 92, 125, 193, 195, 198, 200
workplaces and workers 76–7, 83, 86, 115, 118, 130, 175–7
workshops 56, 65, 83, 108, 125, 167, 178–80, 211

Y

YouTube 83, 85, 87

Z

zines 178, 180n1, 185
zoos *see* museums

www.ingramcontent.com/pod-product-compliance
Lightning Source LLC
Chambersburg PA
CBHW051538020426
42333CB00016B/1996